Human
Ecology

Human Ecology

The story of our place in nature
from prehistory to the present

Bernard Campbell

ALDINE
Publishing Company
New York

First U.S. edition published 1985 by
Aldine Publishing Company
200 Saw Mill River Road
Hawthorne, New York 10532

Library of Congress Cataloging in Publication Data

Campbell, Bernard Grant.
    Human ecology.

    Includes bibliographies and index.
    1. Human ecology. 2. Social evolution. I. Title.
GF41.C355 1984   304.2   84-16733
ISBN 0-202-02025-8
ISBN 0-202-02026-6 (pbk.)

Printed in the United States of America
10 9 8 7 6 5 4 3 2 1

Contents

Acknowledgements

I am deeply indebted to my friend and colleague Sally Rosen Binford for her extensive help and advice in the early stages of the development of this book. Many of the ideas that I have explored were the product of our long discussions. I also wish to thank those who have generously read sections of the book and given me their valuable suggestions: Susan Campbell, W.H. Dowdeswell, Peter Jewell, Colin Leakey, Jim Lewton-Brain, David Wallace, J.S. Weiner, James Woodburn.

I am very grateful for the help and encouragement of my publishers, and especially indebted to Anne Armitage who spent many hours preparing the typescript.

Bernard Campbell

Preface

It has long seemed to me that it is hardly possible for anyone to understand our present evolutionary and ecological crisis without some perspective as to how we came to find ourselves in such grave danger. The idea that we can continually exploit our finite environment to satisfy an ever-expanding population is clearly naïve. Technological advances cannot help us much further. Equally naïve is the idea that we can go back to an idyllic life as hunters and gatherers. What we *should* do to save ourselves is the most important question of this century. I believe that the problem cannot be solved without a thorough understanding of how it was that we came to find ourselves in this terrible predicament.

This short book takes a look at some aspects of humankind's prehistory and history with a view to understanding our evolutionary and cultural adaptations to our different environments. Such a vast subject can necessarily be considered only briefly in such a book, yet I do believe that the principle components of the human adaptation need to be stressed, so that we can examine our present place in nature with a deeper perspective and a more profound understanding.

This book is written for the general reader, because the kind of changes required in our future adaptations to planet Earth are not technological but *political*. It is to be hoped, however, that all biologists, especially those involved in food production, will find the historical perspective of interest and of value.

A people's religion both reflects and determines their attitude to nature. In the first chapter of *Genesis*, verse 27, we read:

> And God said unto them, Be fruitful, and multiply, and replenish the earth, and subdue it; and have dominion over the fish of the sea, and over the fowl of the air, and over every living thing that moveth upon the earth.

In spite of the passage of over 100 years since Charles Darwin wrote the *Origin of Species* (1859), our view of our place in nature, of our relationship to the natural world, is still determined in a profound sense by this early Judaeo-Christian myth. Passages such as this command us to subdue the Earth and all its creatures, and to increase our own population until we come to fill (replenish) the Earth with humankind.

Before the middle of the last century, when a few geologists and biologists came to recognize the great age of the Earth and its inhabitants, it was generally believed (following the calculations of leading theologians) that the creation story described in *Genesis* took place in the year 4004 BC. The passage constitutes an account of humankind's attitude to nature in these earliest years of Jewish history. Abel the pastoralist and Cain the agriculturalist represented two ways of dominating nature which made possible a much higher food extraction rate from the environment than was previously known. These were the two ways of life which were to allow humankind to increase and multiply, and replenish the whole Earth.

The development of pastoralism and agriculture changed in a fundamental way the balance which previously existed between humans and their environment. Not only were certain species domesticated, but others which might compete with domestic herds and damage agricultural land had to be driven from the pastures and fields. Carnivores, which would find the cattle easy prey, had to be kept at a distance or killed. Humankind needed to dominate and control the activity of many mammals with which it began to compete. These revolutionary developments reinforced a trend towards a settled life which had appeared earlier in prehistory, and an increase in population followed.

To the small tribes of nomadic pastoralists and agriculturalists, the Jewish attitude was appropriate. Through these means, humankind has come to dominate and subdue nature and today occupies just about every available suitable area of the Earth's surface. But the small world of the ancient Jews—an area of semi-arid country which includes present day Israel and the neighbouring regions—has now been replaced by a densely peopled Earth. There is neither wilderness to dominate, nor any further space for expansion of human populations. Because the place of the human species in nature has changed, our attitude to it must change as well.

The Judaeo-Christian world view we have described has of course been characteristic of Jews, Christians, and among them especially Protestants. Almost every other religion and philosophy reflects an attitude of greater respect and concern for nature. The importance to us of the Judaeo-Christian viewpoint is not just that it informs much of western technology, but also that in the guise of western commercial values and technology, it is spreading throughout the entire world. A similar attitude of thoughtless exploitation towards the world's natural resources is becoming increasingly common among those from other continents who have been in contact with the West.

In this book an attempt is made to describe briefly the most important human adaptations to different climatic areas of the world: how early humans maintained a natural balance with the animals and plants which constituted their environment, and how with the development of technology and other cultural innovations their rate of extraction of food resources increased. In the later chapters we see how pastoralism and agriculture made possible the exploitation of natural resources in a way that has often been destructive to these very resources themselves. In turn, agriculture has made possible urban development on a vast scale, especially when it is linked with an efficient transport system.

The way the changes have taken place, and the principles which underly them, constitute one of the most significant facts of human prehistory and history. We can describe them with some confidence, because in many parts of the world these different revolutionary changes are rapidly taking place today.

Many of the adaptations to different environments that we shall describe have been critically important in human survival and have influenced human nature in a profound way. Because humankind is a product of its genetic inheritance and its environment, the science of *ecology*—which is concerned with the relationship between a species and its environment—is the key to both an understanding of human evolution and of human nature itself.

We still face the same problems that our ancestors faced—problems of survival. We are still absolutely and finally dependent on the nature of this world, on the plant and animal life with which we live. As Darwin so clearly showed us, we are still part of nature, and our dependence on nature cannot be lessened by our technology nor can that technology ensure our survival. That will depend on one factor only—our success or failure in achieving a new balance with our resources, by stabilizing population growth and the rate of resource extraction. The famine which threatens one third of the world's people demonstrates with terrible clarity that we are already a long way in debt to nature, and her resources are finite.

This book is written in the hope that our vision may be extended a little into the future, by an understanding of our place in nature and by an examination and analysis of our past.

Bernard Campbell
1983

This curious world which we inhabit is more wonderful than it is convenient; more beautiful than it is useful; it is more to be admired than used.

Henry David Thoreau
Commencement address:
Harvard University, 1837.

1 Introduction

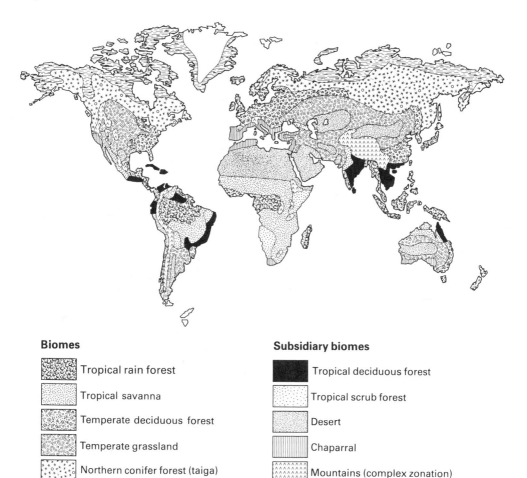

Biomes

- Tropical rain forest
- Tropical savanna
- Temperate deciduous forest
- Temperate grassland
- Northern conifer forest (taiga)
- Tundra

Subsidiary biomes

- Tropical deciduous forest
- Tropical scrub forest
- Desert
- Chaparral
- Mountains (complex zonation)

Fig 1.1 Map of world biomes. The biomes indicated here are fairly stable zones and can be mapped in some detail. Note that only the tundra and northern coniferous forest have some continuity across the northern hemisphere. Other biomes are isolated in separate geographical regions and therefore may be expected to carry ecologically equivalent but taxonomically unrelated species.

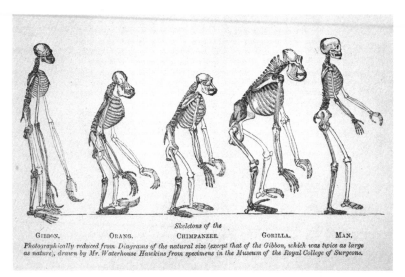

Skeletons of the

GIBBON.   ORANG.   CHIMPANZEE.   GORILLA.   MAN.

Photographically reduced from Diagrams of the natural size (except that of the Gibbon, which was twice as large as nature), drawn by Mr. Waterhouse Hawkins from specimens in the Museum of the Royal College of Surgeons.

Fig 1.2   When the *Origin of Species* was published Darwin was not aware of any fossil evidence for human evolution and did not discuss the question. The evidence from comparative anatomy, however, was very strong and as a result many biologists came to accept the hypothesis of human evolution. Darwin's friend and supporter T. H. Huxley published a book entitled *Man's Place in Nature* in 1863 and in it he wrote 'Whatever part of the animal fabric might be selected for comparison, the lower Apes (monkeys) and the Gorilla would differ more than the Gorilla and Man.' These drawings are done to the same scale except for the Gibbon which is twice natural size. (From T. H. Huxley, 1863)

EVOLUTION AND ENVIRONMENT

Future historians will surely recognize that Charles Darwin's book, published in 1859, was one of the most important ever written. *On the Origin of Species by Means of Natural Selection*[1] not only presented the theory that animal and plant species had evolved over millions of years from relatively simple ancestral forms, but it also descibed an hypothesis which explained how this process actually came about (Fig 1.2). The idea of *organic evolution* or 'the transmutation of species', was not in itself new and has been commonly linked with the names of the French naturalist Lamark as well as with Charles Darwin's grandfather Erasmus Darwin; but the concept of *natural selection*, which explained the process, was original and of extraordinary importance, as it gave the theory a basis—a logical structure which made it increasingly acceptable to those concerned with biology and natural history.

Although Darwin formulated the idea of natural selection first, in 1838, another naturalist, Alfred Russel Wallace, arrived at the same concept independently in 1858. By an extraordinary coincidence, Wallace sent a short account of his ideas to Darwin early in 1858, and as a result their joint paper was presented to the Linnaean Society in London, later that year (Figs 1.3, 1.4).

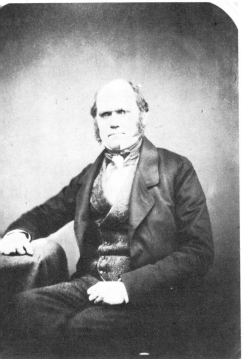 

Fig 1.3 (Left)   Charles Darwin in his 66th year. Darwin lived comfortably at home at Down House in Kent with a substantial private income so that he could devote most of his time to research and writing. An intermittent invalid, he wrote 'Even ill-health, though it annihilated several years of my life, has saved me from the distractions of society and amusement.' He lived to the age of 73. (*National Portrait Gallery*)

Fig 1.4 (Right)   Alfred Russel Wallace, a Welsh botanist, was a complete contrast to Darwin in both background and character. Unlike Darwin, Wallace earned his living by collecting rare tropical plants and animals for private collectors and museums. As a result, he travelled more widely than Darwin in both South America and Southeast Asia. Later in life he wrote a number of books on geography and evolution, but although of considerable interest they do not have the originality and intellectual integrity of Darwin's writings. While Darwin lost his religious belief, Wallace remained a religious man throughout his life. (*National Portrait Gallery*)

Both Darwin and Wallace had travelled widely and observed in great detail the variation that exists within animal and plant species. Members of species, they noted, are not identical, but they vary in size, strength, health, fertility, longevity, behaviour, and many other characteristics. Darwin realized that humans use this natural variation when they selectively breed plants and animals; a breeder allows only particular individuals possessing desired qualities to interbreed.

Both Darwin and Wallace saw that a kind of selection was at work in nature, but they did not know how it worked. An understanding of the means by which selection operates in nature came to both from the same source. The first edition of *An Essay on the Principle of Population* by an English clergyman, T. R. Malthus, had appeared in 1798[2]. In his book, Malthus showed that the reproductive potential of humankind far exceeds the natural resources available to nourish an expanding population.

In a revised version of his essay, published in 1830, Malthus began: 'In taking a view of animated nature we cannot fail to be struck with the prodigious power of increase in plants and animals . . . their natural tendency must be to increase in a geometric ratio—that is, by multiplication.' He continued by pointing out that, in contrast, subsistence can increase only in an arithmetical ratio. 'A slight acquaintance with numbers will shew the immensity of the first power in comparison of the second.' And he had written in 1798, 'By that law of our nature that makes food necessary to the life of man, the effects of these two unequal powers must be kept equal. This implies a strong and constantly operating check on population from the difficulty of subsistence.' As a result he argued that the size of human populations is limited by disease, famine, and war and that, in the absence of 'moral restraint', such factors alone appear to check what would otherwise be a rapid growth in population.

Both Darwin and Wallace read Malthus' essay, and, remarkably, both men recorded in their diaries how they realized that in that book lay the key to understanding the evolutionary process. It was clear that what Malthus had observed among human populations was indeed true for populations of plants and animals: the reproductive potential vastly exceeds the rate necessary to maintain a constant population size. Darwin and Wallace both realized that the individuals that do survive must be in some way better equipped to live in their *environment* than those that do not survive. It follows that in a natural interbreeding population any variation would most likely be preserved, or passed on to future generations, that increased the organism's ability to produce fertile offspring, while the variations that decreased that ability would most likely be eliminated.

Around these ideas Charles Darwin and Alfred Russel Wallace formulated a theory of evolution by natural selection. The theory is not difficult to understand and may be stated as follows:

1 Organisms produce far more offspring than required to maintain their population size, and yet their population size generally remains more or less constant over long periods of time. From this fact, as well as from observation, it seems clear that there is a high rate of mortality among immature individuals.

2 Individuals in any population show much variation, and those that survive do so to a large extent because of their particular characteristics. That is, individuals with certain characteristics can be considered better *adapted* to their particular environment.

3 Since offspring resemble their parents closely, though not exactly, successive generations will maintain and improve on the degree of adaptation by gradual changes in each generation.

This process of variation, and selection by the environment of better-adapted individuals, Darwin called *natural selection*, and the change in the nature of the population that follows upon such selection is the process of *organic evolution*. The same processes occur among both plants and animals. The process of evolution is extremely slow, and to be accepted, the theory required the Earth to be of great age. Darwin and Wallace's theory could not have been accepted by the generations taught by Bishop Ussher, who believed the Earth to be less than 6000 years old. But in his book, *Principles of Geology* (1830-33), Charles Lyell[3] had provided the time dimension required for evolution to work.

The important thing to note here is that the creative process of natural selection is driven by the environment; ultimately by the climatic changes which inevitably occur, and by the immense variety of different environments which the planet Earth carries. Once the process of evolutionary change and the radiation of species is underway, further environmental change due to the appearance of new animals and plants is inevitable, and as the process continues and species multiply, the rate of change tends to accelerate (Fig 1.5). Thus the key to evolutionary

Environment–species feedback loop

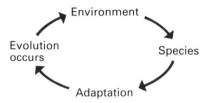

Fig 1.5   Because the environment is acting on all existing species and reducing the
reproductive capacity of those individuals less well adapted, each species evolves,
and in turn brings about changes in the environment of all other species. This is
an example of a positive feedback loop where the processes of change in one
component (the environment) bring about accelerating changes in the system as a
whole.

change is the interaction of the environment and the organisms which
occupy it. The environment, with its ever-changing climatic, mineral,
and organic components, brings about the evolutionary process
through its effect on heritable variation and is a primary factor in the
creation of the multitudinous species of animal and plant life.

Humankind evolved on this planet very recently in geological terms,
and found the world almost in its present condition. Our own evolutionary history, like that of every other organic species, is one of
adaptation to changing environments, but at the same time we have
also adapted to the many different existing environments, and we
occupy a wider variety of them than any other animal or plant species.
An examination of the history of humankind's environments can therefore give us direct insight into the actual process of human evolution.
An understanding of our prehistoric environments can help us understand our own evolutionary adaptations: such knowledge alone can
help us discover why we are what we are—why we are made the way
we are, anatomically and behaviourally.

TERMS AND CONCEPTS

Ecology is the study of the relationship between a species and its total
environment. This is a broad definition, but breadth of scope is the
salient characteristic of the ecological approach. *Human ecology*, then,
refers to the study of all those relationships between people and their
environment (including such factors as climate or soil), and energy
exchanges with other living species, including plants, animals, and

other groups of people. If we take the broadest possible view, human ecology deals with the entire human species and its extraordinarily complex relationship with other organic and inorganic components of the world.

In practice those who have studied human beings from the ecological viewpoint have found it desirable to separate cultural ecology and social ecology as distinct sub-disciplines. *Cultural ecology* is the study of the way the culture of a human group is adapted to the natural resources of the environment, and to the existence of other human groups. *Social ecologists* study the way the social structure of a human group is a product of the group's total environment. In this book we shall consider human ecology primarily in a biological sense, but we shall also attempt to see how human society and culture have developed in response to the environment. We recognize that human group behaviour is dependent upon a body of belief, as well as upon its history, its skills, and its resources. The natural resources of the environment and the skills of the individuals are the primary determinants of human adaptations.

For analytical purposes it is useful to examine units smaller than the entire species; for example to study the ecology of the Eskimos, or African pastoralists, or New Yorkers; when this is done, each human group is treated as a component in a distinct *ecosystem*. The ecosystem is the basic analytical unit of ecology, and it can be defined as any natural association that consists of living organisms and inorganic substances that interact to exchange matter. An example of a natural ecosystem would be a forest or a pond whose animal and plant species depend on each other and on the inorganic chemicals in the environment (Fig 1.6). The classic example of an artificial ecosystem is a fish tank, where the exchanges between plants, fish, their waste products, and the water, are kept in balance with each other. One of the most subtle and complex problems of ecological analysis is to define the boundaries of any ecosystem, and in a sense these are always artificial. Natural systems are always *open systems*—that is, they are to some extent influenced by surrounding systems; however, for analytical purposes they are treated in many respects as though they are *closed systems*; i.e. without reference to related systems.

Human colonization of the world began, not surprisingly, in the tropics and sub-tropics, where there are abundant and varied plants. Only in relatively recent times did it reach the arctic, where humans are dependent upon other predators for food (bears, seals, etc.), and these

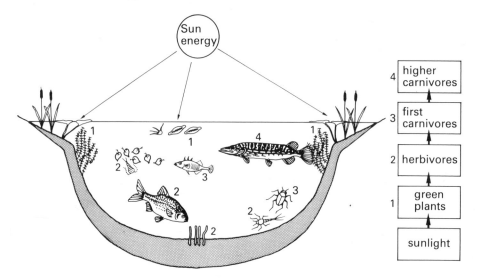

Fig 1.6   Diagram of the ecosystem of a pond. The arrows show the movement of energy from the Sun through the different trophic levels. These are: 1 producers; 2 primary consumers (herbivores); and 3 secondary consumers (carnivores). There are also tertiary consumers (secondary carnivores) and saprotrophs—the bacteria and fungi of decay. The other essential inputs, beside sunlight are the mineral content of the water washed into the pond together with dissolved oxygen and carbon dioxide. (After Odun[4]).

species in turn depend on large numbers of smaller creatures which depend ultimately on plant food (Fig 1.7). The effective hunting and butchering of large animals like bear and seal requires a complex technology and transportation system; since bear and seal are not densely distributed in the environment (it takes a large number of fish to support one bear or seal), humans in the arctic can be supported only in small numbers. In contrast to arctic adaptations, the densest human populations in the world today are in those tropical areas where there is primary dependence on plant resources (Southeast Asia, for example).

The worldwide distribution of the human species is unique among mammals, due to an adaptation peculiar to *Homo sapiens*: culture. By *culture* we mean that system of knowledge, behaviour, and artefacts by which humans cope with the external world. Kinship systems, houses, myths, tools—all fall under the rubric of culture. Culture is of course, the focal point of most anthropological studies, and there are almost as many definitions of culture as there are anthropologists. Many social anthropologists define culture only in terms of ideas and norms held by

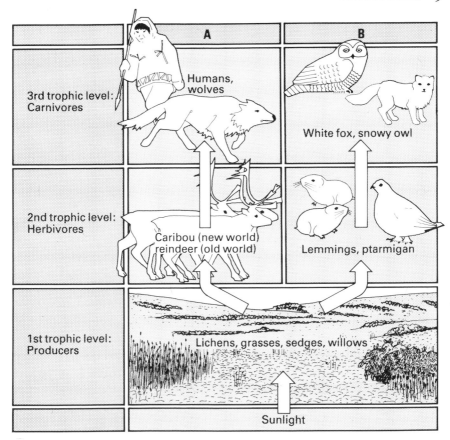

Fig 1.7 Arctic food chains. The Eskimo food chain (A) demonstrates the three main trophic levels and the passage of energy up the chain. Because there is little variety of food in the arctic, oscillations of food supply may be serious (even including marine sources not shown in this chart) and as a result some Eskimo groups have been recorded as practising 'emergency cannibalism' which is not at all common among other hunter-gatherer peoples. The second food chain (B) includes the lemming. Oscillations in lemming populations and their predators are very common and often quite extreme.

a group, but this definition is too limited. In this volume we mean by the term culture all of those ideas, traditions, and artefacts that are acquired by learning. This allows us to distinguish significantly between birds' nests and spider webs on the one hand, and human habitations and hunting traps on the other. The critical point is that the particular kind of nest a bird builds or the kind of web spun by a spider is genetically controlled and cannot be modified quickly to adapt to changing circumstances. Culture on the other hand, allows for much greater flexibility and has the added advantage of being cumulative;

each human generation can communicate a body of knowledge to the next, so that each new generation does not have to learn anew to control fire or invent the wheel.

Thus culture is humankind's learned means of adaptation; it is based on the capacity to communicate symbolically, and it is cumulative. In this sense it is, as far as we know, unique to the human species, although many of our close relatives like chimpanzees and gorillas exhibit kinds of social behaviour that are quite familiar and have been described as *protoculture*. Culture has enabled the human species to spread from tropical to temperate and eventually to arctic regions; to develop a bewildering variety of social customs, and finally to study its place in the ecosystem we call the Earth.

ENVIRONMENT

We have defined ecology as the study of an organism or group of organisms in its relationship to its total environment, and in order to analyse this relationship we must introduce some concepts dealing with the nature of environments. The *environment* of an organism can, as we have seen, be defined as all those objects and forces external to the organism with which it interacts or by which it is affected. In this definition we include other organisms, including other members of the same species. Thus, in human ecology the environment of the species includes other human groups (the *social environment*), plants and animals, climate, natural resources, and so on (Table 1.1).

Environments can be classified according to the attributes they exhibit, and characteristic clusters of attributes—such as mean rainfall,

Table 1.1   Environmental components

Physical factors	Biological factors
Energy	Green plants
Radiation	Non-green plants
Temperature and heat flow	Decomposers
Water	Parasites
Atmospheric gases and wind	Symbionts
Fire	Animals
Gravity	Humans
Topography	
Geologic substratum	
Soil	

temperature, soil conditions, vegetation—that support characteristic forms of animal life. The largest of these environmental units used by ecologists is the *biome*, and some of the most widely recognized biomes (Table 1.2) are shown on the map (Fig 1.1). Only some of the biomes

Table 1.2 Locations and general environmental conditions of terrestrial biomes described in this book (after Billings 1970)[6]

Biome	Principal locations	Precipitation range (mm/year)	Temperature range (°C) (daily maximum and minimum)	Soils
Tropical rain forest	Central America (Atlantic coast) Amazon basin Brazilian coast West African coast Congo basin Malaya East Indies Philippines New Guinea N.E. Australia Pacific islands	1270–12700 Equatorial type: frequent torrential thunderstorms Tradewind type: steady almost daily rains No dry period	Little annual variation Max. 29–35 Min. 18–27 No cold period	Mainly reddish laterites
Tropical savanna	Central America (Pacific coast) Orinoco basin Brazil, S. of Amazon basin N. Central Africa East Africa S. Central Africa Madagascar India S.E. Asia Northern Australia	250–1900 Warm season thunderstorms Almost no rain in cool season Long dry period during low sun	Considerable annual variation; no really cold period *Rainy season (high sun)* Max. 24–32 Min. 18–27 *Dry season (low sun)* Max. 21–32 Min. 13–18 *Dry season (higher sun)* Max. 29–40 Min. 21–27	Some laterites; considerable variety
Temperate grasslands	Central North America Eastern Europe Central and Western Asia Argentina New Zealand	300–2000 Evenly distributed through the year or with a peak in summer Snow in winter	*Winter* Max. −18–29 Min. −28–10 *Summer* Max. 21–49 Min. −1–15	Black prairie soils Chestnut and brown soils Almost all have a lime layer
Temperate deciduous forest	Eastern N. America Western Europe Eastern Asia	630–2300 Evenly distributed through year Droughts rare Some snow	*Winter* Max. −12–21 Min. −29–7 *Summer* Max. 24–38 Min. 15–27	Grey-brown podzolic Red and yellow podzolic
Northern coniferous forest	Northern N. America Northern Europe Northern Asia	400–1000 Evenly distributed Much snow	*Winter* Max. −37–−1 Min. −54–−9 *Summer* Max. 10–21 Min. 7–13	True podzols Bog soils Some permafrost at depth, in places
Arctic tundra	Northern N. America Greenland Northern Eurasia	250–750 Considerable snow	*Winter* Max. −37–−7 Min. −57–−18 *Summer* Max. 2–15 Min. −1–7	Rocky or boggy Much patterned ground Permafrost

shown will be discussed in this book, and zones intermediate between two biomes will sometimes be considered.

The *carrying capacity* of any environment with respect to a given population is defined by the level beyond which no major population increase will occur. We can also calculate the *biomass* for different environments, and this is figured as the sum total of all living matter per unit area, including stored forms of food. There is, of course, a direct and immediate relationship between biomass and carrying capacity.

Some environments, such as tropical rain forests, support many different forms of plant and animal life while others, like the arctic tundra, can support only a few species. The *diversity index* of any community can be expressed as the ratio between the number of species and the number of individuals. Diversity indices (and also density of species) tend to be higher in those areas that are transitional between major biomes. These zones are called *ecotones*, and the fact that ecotones tend to display greater density and diversity than the communities flanking them is known as the *edge effect*. For example, in a zone transitional between a forest and grassland, species will be found that characterize both major environments as well as organisms that are adapted to exploiting the interface between them. In general then, ecotones are characterized by both high densities and great diversity.

Another concept used by ecologists is that of *trophic level*, which is a measure of distance away from the direct utilization of solar energy. Green plants, which utilize and transform solar energy, represent one trophic level. Some animals eat only plants, and such herbivores are all on the same trophic level—the second. Other animals, like wolves, subsist by preying on herbivores, and these carnivores are on the third trophic level in that their food is one step away from direct dependence on plants. Yet other kinds of animals, sea lions for example, eat fish, which in turn eat smaller fish, which depend on yet smaller plant forms. In this case sea lions are on the fourth trophic level. And if humans were present in the environment and hunted the sea lions (as Eskimos do), this would be consumption on the fifth trophic level (Fig 1.7). It is evident that humans, being omnivorous, straddle several trophic levels. There is survival value in such flexibility; an animal that depends on only one or two plants or animals for food is in serious difficulty when this food becomes scarce. We shall return again to this most remarkable aspect of human adaptability—its enormous flexibility.

Types of relationship

Herbivorous ——
Parasitic ··········
Predaceous ━━

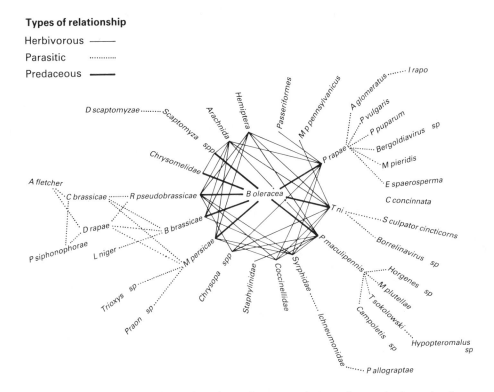

Fig 1.8   If we examine energy flow from the point of view of a particular animal or plant species we soon realize that the food chain is an unrealistic simplification. Most animals consume more than one species of organism and are consumed in turn by many other species. The inter-relationships of organisms as herbivore or carnivorous predator must be supplemented by the phenomenon of parasitism. When we try to analyse the situation in detail we discover, not a linear food chain but a systemic *food web*. The above diagram illustrates only the more abundant 50 species which directly or indirectly derive their energy resources from the cole plant *Brassica oleracea* (after D. Pimental, 1966). Altogether 210 species are involved in this incredibly complex food web. The wide interdependence of these species is clear. The complexity of a food web is such that no one to date has worked out the complete pattern of food relationships in any natural community.

A concept that is closely linked with that of trophic level is that of the *food chain*; this simply describes the paths that energy follows through any ecosystem (Fig 1.8). Ultimately all the energy that reaches the Earth's surface comes from the Sun, and green plants have the capacity to fix and transform solar energy by the process of photosynthesis into their own structural organic materials and into stored sugar and carbohydrates. This vegetation is then eaten, digested, and resynthesized by herbivores, as we have seen, who may in turn be eaten by carnivores or by omnivorous predators like humans. In this way the energy fixed by photosynthesis passes up the food chain and is made available to the entire animal kingdom.

Each life form in a community tends to be found in the same kind of place within the boundaries of the biome, and the kind of place occupied by a species is its *habitat*. Each group of organisms also exhibits a characteristic pattern of exploitation within its habitat—that is, it gets its energy (food) from particular kinds of plants and/or animals in a particular way. This pattern defines the *ecological niche* of a given group of organisms. The ecologist Eugene Odum[4] makes a beautifully clear and simple distinction between habitat and niche—habitat being the organisms's 'address' and niche its 'profession'.

PLAN OF BOOK

Our examination of different ecosystems is organized in order to emphasize several points. As we have seen, ecology is the study of the relationships of matter and energy exchanges in natural systems, and ultimately the source of all energy reaching the Earth's surface is the Sun. Therefore, the degree of *insolation*, the effective solar radiation, reaching any biome should be directly correlated with the biomass of the biome. Tropical regions in fact do have the greatest biomass and the highest diversity indices of all terrestial biomes*, and as we move away from the equator biomes tend to have lower productivity of biomass per unit of area (Table 1.3), and lower diversity indices. These facts, in turn, have had a profound effect on the way in which humans have exploited different environments and have been responsible for structural differences in human ecosystems in various biomes. We shall demonstrate that the greater the distance from tropical latitudes, the longer the food chain, the lower the human densities, and the more complex the technology required for survival. This latter generalization implies that the arctic environment would have been the last habitat to be occupied successfully by humans, and this is, in fact, the case.

We shall begin with a description and analysis of the tropical forest (Chapter 2) and present the evolutionary background of the human species, from a forest-dwelling, arboreal primate to a bipedal, terrestrial form adapted to the tropical forest/savanna ecotone. We shall then follow the earliest hominids—the Australopithecines—into the

---

* Owing to cloud cover over those equatorial regions with high rainfall, the effective insolation is reduced. The highest levels have been recorded in arid tropical regions.

Table 1.3 Net primary production data for the world

Ecosystem	Area 10^8 hectares	Mean net primary production $kJ/m^2/yr$	Total world net primary production $10^9 MJ/yr$
Tropical forest	20	37.80	756
Tropical savanna	15	13.23	198
Temperate deciduous forest	18	24.75	446
Temperate grassland	9	9.45	85
Northern coniferous forest	12	15.12	181
Tundra	8	2.65	21
Agricultural land	14	12.29	172
Open ocean	332	2.42	803

This table shows the approximate area and productivity of the biomes discussed in this book. To these have been added agricultural land and the oceans, for comparison.
kJ = kilojoules; MJ = megajoules. The joule is a unit of energy = 1 watt/second = 10^7 ergs = 0.239 calories. (Data from Odum[4].)

savanna, examine the remains of their and their successors' activities, and attempt to reconstruct their ecosystem as it might have been 2-3 million years ago. For comparison we shall also look at an ethnographically known savanna group, the Hadza, who subsist by hunting and gathering (Chapter 3).

At a later period, humans occupied not only the tropical and subtropical regions of the Old World but also radiated into more northern latitudes. Whether any particular area was temperate or quite cold depended on whether glacial or inter-glacial conditions prevailed at the time of occupation. During the Great Interglacial (approximately 400000-250000 years ago) the area just to the southwest of Peking, China, supported a temperate woodland biome. The cave at Choukoutien has yielded evidence that people occupied the area for an extended period of time, and we shall attempt to delimit the nature of their ecological relations through an examination of the rich archaeological remains recovered from the site. We shall also examine more recent human adaptations to the temperate woodland: American Indian groups of hunters, as they were known at the time of contact with Europeans (Chapter 4).

Generally, a biome known as northern grassland occurs alongside the temperate forest, and they are is bounded on the north by the taiga, or boreal forest. Examination of the nature of the taiga biome in Chapter 5 will indicate why so few human groups occupy this habitat and why, when they do, they always exploit the adjacent regions as well. In prehistoric times the Spanish sites of Torralba and Ambrona were inhabited during a period of glacial cold, about 300 000 years ago, and fossil pollen profiles indicate that some of the human occupations exploited the ecotone between northern grassland and taiga. The excellent data from these occupations provide us with the earliest known exploitation of the taiga. The ethnographic example we shall use to elucidate a form of modern adaptation to the taiga is that of the Tungus—a Siberian reindeer-herding people (Chapter 5).

True tundra has been called a cold desert since it supports so little life, both in terms of numbers of plants and animals, and in variety of species. The first successful occupation of this biome was the Magdelenian, an archaeologically known culture that characterized the terminal Pleistocene of Western Europe (c. 12 000 years ago). An examination of the Magdelenian will be followed by a description of ethnographically known Eskimo groups (Chapter 6).

In our descriptions of human ecosystems from these biomes, both archaeologically known and ethnographically documented examples will be used to illustrate certain ecological principles and to demonstrate that human societies can be viewed fruitfully from an ecological perspective. This perspective must be distinguished from simple *environmental determinism* which asserts that the form and structure of cultures, as well as the psychological capacities of different races of humankind, are directly explicable by reference to a few environmental factors. Moreover, the ecological approach helps us to avoid the pitfall of *psychological reductionism*: this view held that all cultural variations were the result of the psychological preferences of the culture bearers, limited only by historical accident; that there were no regularities in cultural systems, and indeed that culture was not an adaptive system but 'a thing of shreds and patches'[5].

The approach which is adopted here is that human ecology deals primarily with fundamental data about human groups—the question of the food resources of their environment and the extraction of those resources.* Clearly, these basic means by which people get their liveli-

---

* Human ecology also deals with the biological adaptations which different races show to different biomes. These are in general relatively minor factors in climatic adaptation and do

hood are by no means the only determinants of human culture. They constitute data, however, without which no human culture can posssibly be understood. Human culture is the product of human nature, human history, and the human environment: each limits, yet each permits, the growth of individual creativity and the flowering of human society.

Furthermore, it is hoped that by utilizing ecological principles for analysis, we can demonstrate not only that each cultural system functions as the sub-system of an ecosystem, but also that humankind, despite astounding technological advances, are still part of the natural world and risk survival as a species when they ignore this basic fact.

In the final chapters (7-10) we look at some distinct human adaptations: hunting and gathering, pastoralism, agriculture, and urban life. Here we can see in more detail how humankind have exploited their environment through these increasingly powerful adaptations.

Finally (Chapter 11) we take a broader view to see where our adaptations seem to be taking us. There is no doubt that the western lifestyle, which can be characterized by city life based on advanced agricultural technology, gives human beings the possibility of leading fruitful, happy, and rewarding lives. Not everyone, however, can flourish among the varied and often intense stresses of modern life; those who succeed should think themselves fortunate. The price that humankind as a whole pays for this adaptation may be more than the planet can sustain for long; we are living on the Earth's capital—on its non-renewable resources, and the cost to our environment is rapidly becoming irreparable.

References

1 Darwin, C.R. 1859 *On the Origin of Species by Means of Natural Selection* (London: Murray).
2 Malthus, T.R. 1798 *An Essay on the Principle of Population* (London: Murray, 3rd edition 1830).
3 Lyell, C. 1830-33 *Principles of Geology* (London: Murray).
4 Odum, E.P. 1971 *Fundamentals of Ecology* 3rd edition (Philadelphia and London: W.B. Saunders Co.).
5 Lowie, R.H. 1937 *The History of Ethnological Theory* (New York: Rinehart & Co., Inc.).
6 Billings, W.D. 1970 *Plants, Man, and the Ecosystem* 2nd edition (Belmont: Wadworth Publishing Co. Inc.).
7 Weiner, J.S. 1971 *The Natural History of Man* (London: Weidenfeld).

---

not in any major or obvious way influence cultural adaptations or resource extraction capability which is the primary concern of this book. They are discussed in detail by Weiner[7].

2 The Tropical Rain Forest: our distant birthplace

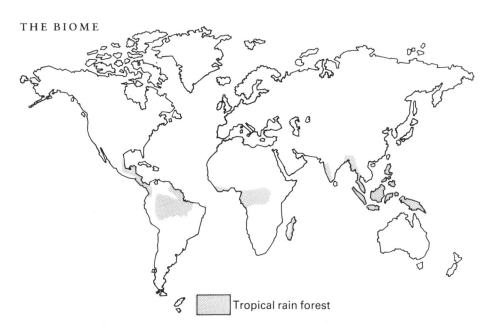

Tropical rain forest

Fig 2.1   The tropical rain forest lies on the equatorial belt, and has developed as a response to a high and regular rainfall. It is organically very rich and diverse in species.

The tropical rain forest was never occupied by early human beings but, like the arctic, was relatively recently entered by people with sophisticated hunting techniques. However, one of the most important results of the investigations which anthropologists have made into human origins is the confirmation of Darwin's deduction that our ancestors originally evolved from forest-living primates of the tropical regions of the Old World. We shall therefore begin this chapter with a consideration of the forest biome and of how those primates most closely related to us, the great apes, have adapted to the forest. We shall briefly discuss their diet and behaviour (as well as that of some monkeys, in passing) as a baseline from which to consider those of our very early ancestors who were so closely related to them, and who shared the rain forest

Fig 2.2   The tropical rain forest of Mayombe in the western Congo where the lush vegetation and dense lower canopy is interspersed by immensely tall evergreen trees. (*Topham*)

with them. The primates, an order of mammals which includes humans, apes, and monkeys, as well as a number of less well-known groups, is characterized above all by a whole range of adaptations to arboreal life.

The tropical forests of Africa, America, and Asia are the most productive biomes in the world. Here the variety of plant life reaches its maximum; rainfall is heavy (usually over 2500 mm per annum) and well distributed around the year; seasonal variation in temperature is minimal and less than the daily variation (21–32° C). Tall evergreen trees dominate the forest, and form the upper story of the habitat, growing to a height of 35–45 metres, while a few trees sometimes reach a greater height. Below these grow the middle story, and they occupy the quite extensive spaces between the trees of the upper layer. Below them in the remaining spaces there is an almost continuous layer of smaller trees from 8–15 metres in height (Fig 2.3). These carry a dense mass of woody creepers and epiphytes, and utilize the remainder of the sunlight that penetrates the higher levels. As a result of this dense multiple canopy, the forest floor carries only a poor cover of herbaceous vegetation, and parts of it may be bare. But the high level of insolation

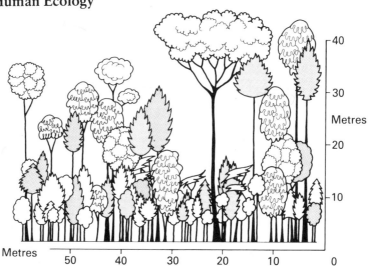

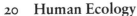

Fig 2.3 A profile diagram of an actual strip of tropical rain forest in Nigeria shows the height of the canopy and its stratification. There are three main areas of tropical rain forest: the Amazon and Orinoco basins and Central American Isthmus; the Congo, Niger, and Zambesi basins in Central and West Africa, and parts of Madagascar; and regions in Indonesia, Malaya, Borneo, and New Guinea. All these forests carry a number of species of monkeys, and in the Old World, apes. Only lemurs occur in Madagascar.

and the high and regular rainfall mean that the tropical rain forest is one of the richest biomes in the world in terms of its diversity, productivity, and potentially edible organic material.

THE AFRICAN APES

The present evidence suggests that the ancestral mammals which evolved into the primates were small ground-living animals closely related to the present-day shrews. Living on a diet of insects, eggs, and possibly some soft vegetable matter, these creatures exploited the forest floor, and probably moved up the trunks of trees in their search for insect life. So long as they remained small, they must have paralleled to some extent the squirrel's adaptation which we see today, but they occupied a different niche in the forest. While in modern temperate forests squirrels live exclusively on buds, nuts, and seeds, the early primates were insectivorous and only slowly became adapted to a diet with an emphasis on vegetable matter.

The change in emphasis of the primate diet in a forest environment was the key to the evolution of the monkeys and apes. While most species of the higher primates remained omnivorous, but got the majority of their calorific intake from forest vegetation, a major group

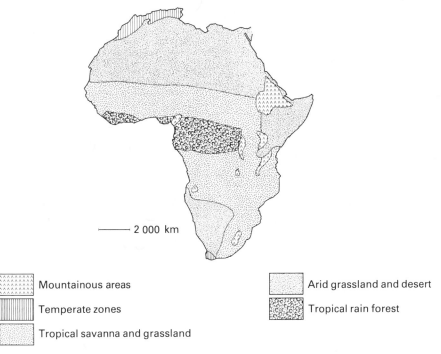

Mountainous areas

Temperate zones

Tropical savanna and grassland

Arid grassland and desert

Tropical rain forest

Fig 2.4 The African biomes centre on the tropical rain forest and its surrounding woodland and woodland savanna, which cover a vast area of very rich hunting grounds. In early Miocene times the rain forest may have extended to the East coast. (After Odum)

of monkeys, the Colobinae, became entirely herbivorous and developed an enlarged, sacculated, and folded stomach to assist digestion of cellulose, rather like ruminants. The fact that not all the arboreal primates adapted to a fully herbivorous diet is extremely important: by retaining a taste for meat, the apes could give rise to a balanced omnivore such as *Homo sapiens*. A mixed diet is therefore characteristic of most other primates, both the Old World monkeys and the apes. Such an adaptation means that these primate species remained adaptable and were therefore able to change biomes when the conditions were right. Well known examples include the baboons, many kinds of which now live in open grassland with few or no trees. While primarily vegetarian, they do kill small mammals and have been seen to hunt in groups on occasion (Fig 2.5). A number of other species of monkey live in open country and have adapted their diet accordingly.

Of the two great African apes, the gorilla, which has been studied in some detail by Schaller[1] and Fossey[2], eats a very varied vegetable diet:

Fig 2.5 Baboon diet consists almost entirely of grass shoots, but under certain conditions baboons will readily take to eating meat and even hunting. Many higher primates have this potential to switch to an omnivorous diet. (*Shirley Strum*)

Fig 2.6 Chimpanzees in the Gombe Stream Reserve in Tanzania enjoy intermittent hunting and meat eating. Meat constitutes no more than 5% of their diet but is greatly appreciated. Here, they are sharing the remains of a red colobus monkey. (*Jane Goodall*)

mostly succulent plants, or the pith and bark of herbs, vines, and bamboo according to altitude and season. Schaller recorded 100 different food plants in the area which he studied, but fruit did not constitute a significant part of their diet. Studies of chimpanzees gave rather different results. Reynolds[3] reports that in Uganda fruit constitutes 90% of the total bulk consumed, leaves and bark some 9%, and insects only 1%. The chimpanzees of the more open woodland which Jane Goodall studied in Tanzania[4] had a more varied diet, and while they were mainly frugivorous, they also fed regularly on ants and termites, caterpillars, honey, birds and their eggs, and from time to time on mammal meat (Fig 2.6). On a number of occasions chimpanzees were seen to catch and tear apart young monkeys which they ate with gusto. Evidence from their faeces shows that they also ate the red colobus monkey, blue monkey, red tail monkey, bushbuck, and bushpig. Hunting, however, is only rarely observed, but Goodall has described how a group of male chimpanzees caught a colobus monkey by a combination of surprise and speed. In general, chimpanzees seem adaptable in their diet, which is very diverse, and this adaptability is also seen in gorillas which take readily to meat in captivity. We share with these apes this capacity for an omnivorous diet.

We need to look beyond the diet itself, however, to consider how primate populations are distributed with respect to food supply and territory. This leads us to consider the forest as an ecosystem in equilibrium, of which the primates are one component; and to consider how natural selection has brought about a stable balance between primates, other animal species, the trees, and other food resources.

It has often been assumed that the overall population density of a species is limited by its food supply, but careful investigation of a number of species has demonstrated that factors other than food usually limit population. The only primates that have been investigated from this point of view are the savanna-living baboons and in their case the limiting factor has been shown to be the availability of sleeping trees or cliffs. We do not know enough of the biology of forest-living primates to pin-point the causes of known densities: there is no direct and simple relationship between population and either food supply or any other single aspect of the environment. Densities among non-human primates vary from 100 per 2·5 km² (square mile) among howler monkeys, to between 3 and 10 per 2·5 km² among chimpanzees and down to one per 2·5 km² among gorillas. Chimpanzee gestation is 33 weeks, ovulation ceases after giving birth and does not begin again until about three years later when the infant is weaned. At the same time there may be a mortality rate among infants and juveniles of about 50%[4]. The size of the population may oscillate a little as it does in many mammals, but this occurs well below the point of exhausting food resources.

One important factor in primate biology which *is* limited by food supply is the formation and size of social groups, or troops. Troop formation is made possible by a certain density of food resources. All Old World higher primates live in troops (except one langur, the orangutan, and the gibbon) but those species which are ground-adapted and live in sub-desert regions such as the Hamadryas baboon break up into small 'family' groups, each dominated by a male, during the day when they disperse over a wide area to feed on the sparse and scattered resources. Where the food density allows permanent troop formation, because it permits them all to feed in a limited area, the primates reap certain very clear benefits. One of the most valuable features is the possibility of one individual learning from an older more experienced one, be it parent or peer, by observation. Any arrangement which facilitates the transmission of useful behaviour is advantageous to a species which depends on learning to develop its full behavioural repertory. A second advantage of social grouping is that it affords protection of the females and young of the group by the males. This is exceptionally important where relatively small mammals face powerful predators. As protectors, the males are expendable, for a few can fertilize many females.

We can see that the existence of the social group is made possible by a certain richness in the availability of food. The troop already assumes a number of important functions in forest-living primates. Among those adapted to the forest/savanna ecotone and the savanna itself, the functions of social groups are, as we shall see, multiplied. In the evolution of human beings, social adaptations become as important as biological adaptations, and form an integral part of our adaptation to our environment.

THE FOREST/WOODLAND ECOTONE AND BIPEDALISM

Along the borders of the Miocene rain forest stretched vast areas of woodland, as they still do today (Fig 2.7). This is country where the rainfall is lower and irregular and it carries only a single closed canopy of trees, which are often deciduous. Grasses and herbs constitute the dominant ground cover. Such tropical woodland is an important

Fig 2.7 Tropical woodlands contrast with the rain forest, with their small trees and lower rainfall. Chimpanzees have been studied extensively in this more open environment with its very diverse forest-edge fauna and flora. Here, Jane Goodall searches the woodlands of the Gombe Stream Reserve for chimpanzees. (*Jane Goodall*)

feature of both Africa and Asia, and as we shall see, it may have played a critical part in the early evolution of the Hominidae. It is of especial interest that the chimpanzees studied by Jane Goodall, mentioned earlier, are found in a woodland area. Although they are the same species as the rain forest chimpanzees found in Central and West Africa, the woodland populations are believed to consume far more mammal meat, hunt more often, and modify objects for use as tools more often than their cousins in the rain forest. Because the ground cover is grass and herbs, the woodland carries a very different fauna to the dense rain forest, and of course it is more easily penetrated by ground-living herbivores and carnivores.

The present evidence for the course of human evolution suggests that the early *hominids* (animals which belong to the family of humankind and which evolved into modern people) evolved from ape-like creatures which first occupied the tropical rain forest and later spread into the associated woodland. The forest edge is made very extensive by natural irregularities of altitude and rainfall, and by the presence of forest along the river beds which cross the more open plains. These riverine forests are somewhat different in nature to the rain forest, but are no less productive, nor are they less rich in primates. Their long narrow shape means that no part of the forest is far removed from woodland, and there is great extension of the forest/woodland ecotone.

Fig 2.8   A possible human ancestor is represented by this small fossilized jaw from Fort Ternan in Kenya dated at about 14 million years BP. The canine teeth are significantly smaller (relative to the other teeth) than those seen in living apes.

The fossil evidence which we believe may represent this early stage in human evolution consists of jaw fragments and teeth from East Africa and Turkey dated at about fourteen million years before present (Fig 2.8). On the basis of the other fossil animals associated with them (most of which are today forest living), these creatures are believed to have lived in riverine or montane forest near to a forest/woodland ecotone.

Fossil apes from a later period, dated from about nine to eight million years ago, have been found in Greece, Turkey, Pakistan, and India. These are associated with a fauna which includes animals characteristic of woodland, and even the more open savanna grassland. Indeed the evidence from many other sites suggests a profound change in the climate throughout North Africa, South Europe, and Asia at about this time, so that by five million years ago there was no rain forest left in Southern Europe and central and eastern Asia, nor were there very large areas of woodland. Most Miocene apes had died out with the forest by about eight million years ago, and the only creatures to survive were those which had successfully adapted to woodland, and even to the more open savanna.

To understand the significance of those fossil apes which may have given rise to the earliest hominids, we need to know something of the

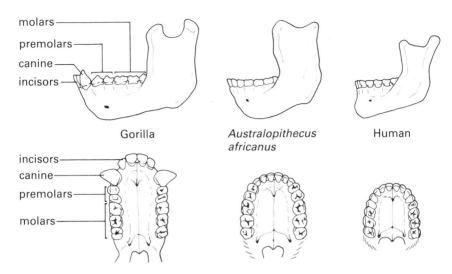

Fig 2.9   These drawings show the palate and a lateral view of the jaws of a gorilla, *Australopithecus*, and a modern human. The striking difference in the size and form of the canine teeth is clear, as are differences in shape of the dental arcade. In the ape jaw the molar teeth are adapted for crushing, in the human and *Australopithecus* jaws, for grinding.

function of the teeth of non-human higher primates, especially the living apes. They are adapted for cutting (incisors), for tearing (canines), and for chewing (premolars and molars). In the Eurasian fossil apes we find a relative reduction in the size of the incisors and canines (compared with the living and other fossil apes) and a relative increase in the size and surface area of the premolars and molars. In particular, we find that these molars are worn rapidly, and from this we deduce that they were used for grinding as well as chewing. This trend towards 'molarization' we also find in the later hominid fossils: it suggests a change in diet towards dependence on tougher plant foods such as grass, berries, seeds, and roots which would have carried grit. The reduction in size of incisors and canines suggests a dependence on tools for cutting and tearing (Fig 2.9). From Fort Ternan in Kenya, where some of these fossils were found, we have evidence that stones may have been used for smashing bones. It is also reasonable to infer that materials other than stone (especially bone and wood) would have been used at this time as tools. We can be fairly sure that hominid evolution first involved adaptations to the forest floor, followed by successful adaptations to more open woodland, and eventually to savanna.

A species living at the junction of two biomes is clearly in a position to exploit a double set of resources. During the period when food is seasonally at its minimum in the forest, such a species would be in a position to find an alternative source in the woodland, and *vice versa*. For this reason, an ecotone is usually rich in fauna, and the forest/woodland ecotone in Africa today is rich in monkeys, both as regards species and numbers.

Beside the differences in seasonality, the food of the woodlands differs in a number of ways. The dry seasons have brought about the evolution of plants which resist desiccation by various means, and this results in their increased toughness. We also find the evolution of food-storage organs which carry the plants through the drought. These organs (usually seeds or tubers) are very rich in starch, and also contain some protein: they have a high nutritional value. The exploitation of toughened vegetation containing silica, protected seeds, and tubers covered in grit, would have been a significant factor in the adaptation of the early hominids to more open country. The increased wear on the molar and premolar teeth of some fossil apes and their large size and thick enamel all bear witness to this development. The success of this adaptation accompanied eventually by a greater dependence on meat, also owes much to the unspecialized nature of the primate digestive system.

One of the most important adaptations of the early hominids was of course the bipedal run and walk. We have no fossil yet from the earliest period which can tell us of this development, but we do know that by the time of *Australopithecus* (c. 4 million years BP), bipedalism was already fully evolved. It is likely that during the preceding millenia a slow change was taking place in the spectrum of locomotor behaviour, so that greater reliance was placed on bipedalism, and less on quadrupedalism. Observations of chimpanzees, which spend some part of each day on the ground and often move bipedally, have given us some ideas of the advantages and disadvantages of bipedalism. The advantage of freeing the hands from full-time locomotion would be great. Not only could food and tools be carried, but the hand could evolve as a precision organ, with precise and skilful muscular control: a prerequisite for tool-making.

A second advantage would be gained from improved vision. Ground-living primates are prey to big cats and other powerful predators and they depend on sight (rather than scent) for warning of danger. The superb eyesight which evolved as an adaptation to arboreal life was no doubt of inestimable value. Primate eyes are better in all respects than those of other mammals, being capable of excellent definition and stereoscopic colour vision. The advantage of eyesight over smell is that it is multi-directional: it does not depend on the direction of the wind. By standing erect primates gained valuable information about quite distant animals; some were predators to be avoided, and some were prey.

For escape, primates have (with very few exceptions) only one solution: to climb a tree. All ground-living species sleep in trees (or on cliffs) and require trees as an escape from danger. Reliance on trees limited the dispersal of the earliest hominids into grassland savanna plains, since they had no large canines for defence, and were in any case, relatively small animals. It seems clear then that these creatures were confined to woodland savanna until a late stage in their evolution. Only with defensive weapons or fire would it have been safe for hominids to leave the vicinity of trees, especially at night.

Thus both the fossil evidence and theoretical considerations point to the forest/woodland ecotone as the environment in which the early hominids evolved. Here alone, intermittent bipedalism would have been advantageous, while the protection of the trees was still available. Here the early hominids found an enriched food supply, and their primate vision could be put to good use in the detection of danger, and the identification of animals to prey upon.

RETURN TO THE FOREST

It was not until relatively recent times that human technology enabled people to re-enter the tropical forests. Today, forest-living peoples are found in the rain forests of Africa, Southeast Asia, and South America. In almost all instances, however, these people have adopted horticulture, often quite recently, and though they may still hunt, in only a few instances do they live by hunting and gathering wild foods alone. The earliest evidence of horticulture, of forest clearance, comes from central Africa and dates from about 3000 BP.

In the Congo basin of central Africa, we find people well adapted to life in the rain forest. These people are pygmies, some 18–25 cm shorter in mean stature than Europeans, and those we know best are the Mbuti of the northeast region of the Congo forest[5] (Fig 2.10). Although these people often trade with village cultivators, they are able to derive their own living from the rain forest, and their forest adaptation is expressed clearly in their diet and technology. The Mbuti bands (which consist of

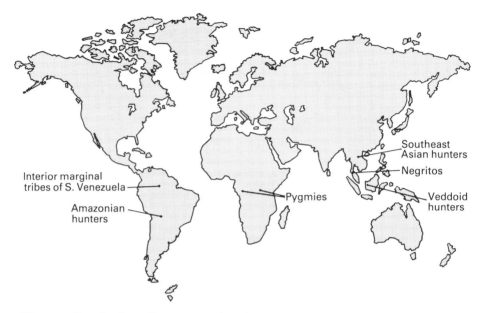

Fig 2.10   People of small stature are found in tropical regions throughout the Old
    World and are associated with tropical rain forest. This map shows the homes of
    some of the best known groups.

7 to 30 families) define their hunting territories but do not defend them.
They are often extensive and may amount to as much as 500 square
miles for a band of from 50 to 150 people. The density in the forest is
therefore extremely low (about 0·25 people per km²) and famine is
unknown to the Mbuti[5-6].

The low density of forest hunters and their relatively recent arrival
in the rain forests are probably due to the fact that the rain forest carries
a relatively small amount of game, and much of that is in the canopy in
the form of birds, monkeys, and other arboreal animals. While something like 50% of grassland production may be exploited by grazing
mammals, probably not more than 2·5% of rain forest production is
similarly exploited. It follows that meat is not plentiful in rain forest,
and since, as we shall see, humankind's ancestors have most probably
usually consumed a significant percentage of meat, it would appear
that this factor, together with the difficulty of hunting in a rain forest,
may have kept humankind out of the forest until, perhaps, the last
20000 years. This figure, however, is no more than a guess.

The Mbuti settle in forest clearings, and build their houses of saplings
and leaves (Fig 2.11). They may remain in one place for a month or two
until food resources become scarce, and then move on. In common

Fig 2.11 A Mbuti forest camp. The Mbuti people's social organization, lifestyle, and technology is fully adapted to the rain forest—one of the most difficult biomes for humans and one of the last to be entered. (*Ed Tronick, Anthro-Photo*)

Fig 2.12 A Mbuti woman shells peanuts traded by the Balese – neighbouring agriculturalists. (*Nadine Peacock, Anthro-Photo*)

with most hunters and gatherers these people live primarily on vegetable foods collected by the women—fruit, mushrooms, nuts, and roots—which form about 70% of their diet. Although hunting is a more prestigious activity, food gathering strategies play a large part in determining their daily and seasonal movements. The supply of vegetable food is constant so there is no need to store food. In hunting, the Mbuti employ two techniques: the bow and arrow, and the net and spear. The bow and arrow hunters live in small groups of a few families—perhaps half a dozen—and hunt birds and mammals with poisoned arrows. They may kill animals as large as buffalo, hippopotamus, and even elephant (though these, like their hunters, are smaller in size than their plains-living relatives) as well as antelope and pig. The net hunters operate in larger groups of up to 30 families. Large bands are demanded by the technique: the women drive the forest floor game into the family-owned nets, which are strung together to form a wide semicircle (30 to 90 m wide by 1 m in height), and here the men spear them. Small game is consumed by the owner of the net into which it runs, but large kills are shared between families. It has been estimated that among net hunters each family gets the equivalent of half an antelope every day. Water presents no problem, as streams are abundant and flow throughout the year.

In the absence of seasonality in either rainfall or insolation, the Mbuti emphasize the honey season as an annual event of importance. At this time the small bands of archers gather and social bonds between families and bands are re-established. Thus the integrity of the larger breeding group is maintained in spite of the need for a wide distribution of bands throughout most of the year to exploit the uniformly distributed food resources. (The social bonds which always exist between human bands contrast clearly with the total independence of non-human primate troops.)

Today, the Mbuti trade quite extensively with the village cultivators (Fig 2.12). They supply building materials, meat, and honey from the forest, as well as some service in the plantations, in exchange for products of the plantations such as manioc, rice, beans, and bananas, as well as tobacco and palm wine. They are wholly dependent on the villagers for metal goods (especially knives, axe-heads, and arrowheads), as well as pottery. None of these imports, however, appears to be essential: the poisoned arrow tip is wooden, and the nets and spears are made from forest products. Today, however, the Mbuti are certainly becoming somewhat dependent on their trade, yet there is no doubt of their ability to survive without it. It seems clear that these pre-agricultural people have until very recently survived as an independent forest group: as part of a stable forest ecosystem.

In Southeast Asia, we find people adapted to the rain forest who are also of small stature. One of the groups that we know best is the Semang of Malaya who are sometimes described as Negritos[7]. These and other somewhat similar people (such as the now almost extinct Andaman Islanders and Philippine Negritos) also show true rain forest adaptations. The Semang subsist on shoots, leaves, nuts, berries, fruits, and roots, and on small game such as rats, squirrels, birds, and lizards, together with occasional pigs, which they kill with poisoned arrows. Fish are an important source of additional protein in some areas. These people show many parallels of adaptation to those of the Congo basin, but the most striking is probably their small stature.

It is clear that today all forest-living people are in possession of a well-developed hunting technology, with bows, poisoned arrows, nets, and spears. In other respects their technology is fairly simple: having evolved in the tropics, humans do not require protection from this equable environment. As we have seen, the Mbuti live in bands of at least a few families (perhaps 12–15 people), and their seasonal movements in the forest follow the need to gather plant food and to find game. The small bands gather annually into larger groups, and this presumably serves to maintain a breeding population. The people as a whole are well nourished and their numbers stable: it seems clear that as far as food resources are concerned the forest could possibly support twice the population. Turnbull maintains that the forest population is maintained at its present level for political reasons: to avoid conflict with neighbouring groups. From what we know of the potential fertility of human populations, we must suppose that as well as disease, and the natural inhibition of ovulation during lactation, some behavioural mechanism operates to lower the birthrate. This may take the form of taboos against sexual intercourse during certain periods, or be brought about by abortificants, or infanticide, all of which are known among hunter-gatherers.

The pygmies exploit a wide range of food resources in the forest (with its high diversity index) across two trophic levels. This contrasts markedly with the much less reliable, because less diverse, production of vegetables by horticulture and agriculture. In those areas where hunters have made contact with farmers and have adopted a new subsistence pattern, they survive times of food shortage by returning to

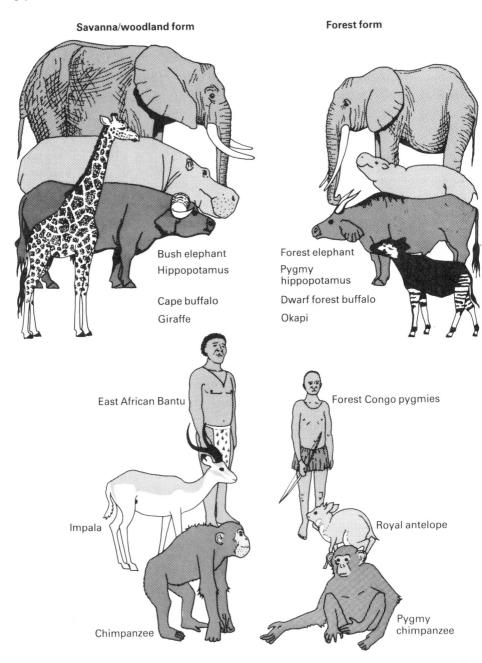

Savanna/woodland form

Forest form

Bush elephant
Hippopotamus

Forest elephant
Pygmy hippopotamus

Cape buffalo
Giraffe

Dwarf forest buffalo
Okapi

East African Bantu

Forest Congo pygmies

Impala

Royal antelope

Chimpanzee

Pygmy chimpanzee

Fig 2.13 Many different animal besides humans have adapted to the forest through a reduction in size. Some of the most important are shown here against the more common and larger savanna variety.

the forest to hunt. The village cultivators in the Congo basin depend to a considerable extent on the neighbouring pygmy hunters for their meat supplies. In South America, some tribes show a recent reversal to the more reliable hunting economy, and no longer practice agriculture, for which the forest soils are not well suited[8].

SUMMARY

In this chapter we have seen that the richness and variety of food resources in the tropical forest have allowed the evolution of highly social primates which live in troops and bands. The diversity of species makes a reliable year-round food source for consumers. There are rarely less than 12–15 individuals in a group. The factors which limit population density are not known in either example, but in both behavioural spacing mechanisms stabilize the density and size of populations in accordance with the area of forest available. There is no evidence of large oscillations in population densities. Distinct races of humans, chimpanzees, and some other large animals (such as elephant and buffalo) have become adapted to the rain forest and are characterized by small stature (Fig 2.13).

The central differences between the ancestral and descendant human primates of the rain forest arise from humankind's commitment to terrestrial life and the dietary consequences. Unable to exploit the forest canopy, like the chimpanzee, people have expanded the range of vegetable foods consumed to include mushrooms and roots, and added a whole new trophic level in the form of game, which does not constitute a significant part of the chimpanzee's diet. The pygmies are said to recognize the human-like form of the chimpanzee, and they do not place him among their game animals. The two species appear to co-exist without interaction: they share a habitat, but occupy different niches. Today, gorillas, chimpanzees, and forest-living humans are few in number. As we shall see in the last chapters, the development of agriculture, and the exploitation of forest timbers throughout the world, threatens the whole biome, together with its fauna and humans.

In the following chapters, we shall discuss humankind's adaptations to the savanna grasslands of Africa, and the cooler biomes of the northern hemisphere. As we move away from the biotic riches of the forest and the tropical environments, we shall see that people become increasingly dependent on the evolution of complex social behaviour and the development of technology.

References

1   Schaller, G. B. 1965 *The Mountain Gorilla: Ecology and Behaviour* (Chicago: University of Chicago Press).
2   Fossey, D. 1976 *The Behaviour of the Mountain Gorilla*, PhD thesis, Cambridge University.
3   Reynolds, V. 1965 *Budongo: an African Forest and its Chimpanzees* (New York: Natural History Press).
4   Goodall, J. 1969 The Behaviour of Free-living Chimpanzees in the Gombe Stream Reserve. *Animal Behaviour Monographs* 1:161-311.
5   Turnball, C. M. 1965 *Wayward Servants* (New York: Natural History Press).
6   Schebesta, P. 1933 *Among Congo Pygmies* (London: Hutchinson).
7   Schebesta, P. 1929 *Among the Forest Dwarfs of Malaya* (London: Hutchinson).
8   Lathrap, D. W. 1968 The 'hunting' economies of the tropical forest zone of South America: an attempt at historical perspective. In *Man the Hunter* R. B. Lee and I. DeVore, eds. (Chicago: Aldine Publishing Co.).

3 The Tropical Savanna

Tropical grassland and savanna

Fig 3.1 One of the most important biomes in the story of human evolution, the tropical grassland covers vast areas adjacent to tropical forest, and is a response to lower and more seasonal rainfall.

One of the most spectacular of the Earth's biomes, both in terms of its vast extent as well as in the number of herds of large animals it supports, is the tropical savanna (Fig 3.1). While South America and Australia contain large expanses of tropical grassland, the best known and the most relevant for our purposes is the tropical savanna of Africa, and especially East Africa (Fig 3.2). It was here, as far as we know, that our Pliocene ancestors developed a way of life that demanded behavioural adaptations which we consider uniquely human; it was in adapting to the tropical savanna that humankind developed its peculiar posture and gait, its eclectic omnivorous habits, and probably its

Fig 3.2   The East African savanna is famous for its vast plains and huge herds of
herbivores – mostly different species of antelope. This shows a typical savanna
landscape, and part of a herd of wildebeest or gnu. (*Frank Lane Agency*)

capacity for abstract symbolization that distinguish it from its close
cousins, the non-human higher primates.

The term savanna is today often used colloquially to refer to the
open grasslands such as we find in the Serengeti plains of East Africa,
but the term is more properly used to refer to very open deciduous
woodland which blends with the denser woodlands described in the
previous chapter surrounding the tropical rain forest. It is widely held
that savanna grassland without trees is a geologically recent result of
burning and overgrazing which, as we shall see, is characteristic of
much of this area in East Africa. However, savanna is usually characterized by a certain amount of tree cover, from quite large deciduous
trees, such as baobabs, to small acacias, and in the driest areas (sometimes called bushland) it carries a fairly dense cover of bushes and small
bushy trees. The latter type, which has much less grass, is often a
product of overgrazing.

It is believed that the family of plants called grasses (Graminae)
evolved about 50 million years ago, and that the savannas began their
development some 25 million years ago. The appearance of these
grass-covered plains was due to changes in rainfall pattern, which
caused the development of seasonal wet and dry periods, with an

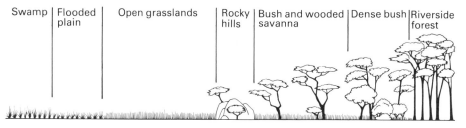

| Swamp | Flooded plain | Open grasslands | Rocky hills | Bush and wooded savanna | Dense bush | Riverside forest |

Fig 3.3   The savanna is characterized by extensive areas of grazing interrupted by a number of other features. It is often intersected by riverside forest, and contains clumps of denser bush, rocky outcrops, and occasional swamps.

annual rainfall of 800–1600 mm. Animals moved out from the previously extensive forests and woodlands and began to adapt to this new and remarkable biome. The grasses which constituted most of the vegetation presented an unusual and productive source of food (Fig 3.3). With the plants generating their leaves from buds at or below ground level, the leaves could be cropped by grazing animals without the stems being damaged, which could thus produce leaves continuously. With stems below ground, grasses could also survive the dry seasons typical of savannas, and the fires that often raged across the plains. They also contained stiffening crystals of silica and thick cellulose cell walls which helped to protect them from collapse in drought. Though they found their food far easier to obtain, the animals which adapted to eating grass required much tougher grinding teeth than were necessary in the forest. This diet also selected the ruminant adaptations for the digestion of cellulose, with the help of an enriched intestinal fauna and flora. In order to give space for the slow fermentation process, the digestive tract lengthened, extra lobes of the stomach evolved, and 'chewing the cud' appeared among many Bovidae.

The most striking aspect of life on the savanna is the large number of herds of ungulates that can be supported: in Africa the characteristic fauna includes antelopes, zebra, giraffe, and the predators they support, such as lions and hyenas. The numerous species of antelope, together with the other grazing and browsing mammals of the savanna, today more numerous and various than in any other place on Earth, were the first large animals hunted by our distant ancestors. But in order to understand the success of the human species as a hunter (which enabled it to inhabit almost every part of Africa, Europe, and Asia long before becoming dependent upon domesticated animals and food crops), we must look to the origin of this behaviour. In this chapter we shall examine the adaptation to the tropical savanna of two very different kinds of hominids: the australopithecines, a term including hominids of Pliocene age and their successors, the earliest humans; and for comparison a modern tribe who also have made their living hunting and gathering in the East African savanna, the Hadza.

AUSTRALOPITHECUS — EARLY WORK

When Raymond Dart of Witwatersrand University, Johannesburg, found the first remains of a juvenile australopithecine in 1924 and

Fig 3.4   In 1924 Raymond Dart startled the anthropological world by his discovery of a small fossil skull from a quarry in Taung in South Africa. He named it *Australopithecus* and declared it to be a human ancestor. His claim was derided, but finds made many years later proved that his fossil was indeed hominid. Here, Raymond Dart shows the skull to Louis Leakey. (*Alun R. Hughes*)

Fig 3.5   The magnificent skull of *Australopithecus africanus* found at Sterkfontein
by Robert Broom in 1947. Although the teeth and jawbone are missing, the skull
itself is undistorted—a rare find. (*D. Panagos, Department of Palaeontology,
Transvaal Museum*)

interpreted it as representing a population ancestral to man, the disbelief of his colleagues was almost universal (Fig 3.4). During the period
from 1935 to 1955 Dart, Broom, and others in South Africa accumulated
an impressive number of australopithecines from cave deposits, and a
major reinterpretation of the hominid fossil record became necessary.
On the basis of the skull's positioning with respect to the spinal column,
Dart asserted that the australopithecines were upright and probably
bipedal, a claim that received scant support from the majority of
anthropologists. Later finds confirmed Dart's early interpretation;
pelves and other post-cranial remains indicated clearly that this relatively small creature habitually moved in a bipedal manner. Dart also
identified pieces of worked bone in the cave deposits that yielded the
australopithecines, and postulated that these were deliberately shaped
as tools. Since the mean cranial capacity of the australopithecines was
about 500 cc (*versus* a 1350 cc mean capacity in modern humans), this
interpretation was also met with wide disbelief. Tool-making, it was
thought, was a uniquely human capacity and was therefore dependent

upon the possession of a brain considerably larger than that of the australopithecines. The early 1950's was a period of conflict in interpretation of both the fossil record of these rather unimpressive creatures and their association with 'tools'. Since South Africa yielded the only australopithecine remains and since these were recovered as a by-product of industrial mining of lime from consolidated cave deposits, the archaeological associations were not firmly established and interpretations remained essentially expressions of opinion (Fig 3.5).

AUSTRALOPITHECUS AND EARLY HOMO:
RECENT DISCOVERIES

The general scepticism about the tool-making capacities of early hominids was shaken by the find in the mid-1950's of stone artefacts in a layer of one of the South African caves that also yielded hominid teeth. Since the late 1950's, our understanding of the significance of these early hominids has grown considerably.

Evidence is accumulating rapidly that *Australopithecus* occurred over wide areas of Africa during the last five million years. Sites are now known not only from South Africa but also from the southwestern and northeastern regions of Ethiopia, from northern Kenya, and Olduvai Gorge and Laetoli in Tanzania (Fig 3.6). The remarkable discoveries of Don Johanson and his team in northeast Ethiopia, of

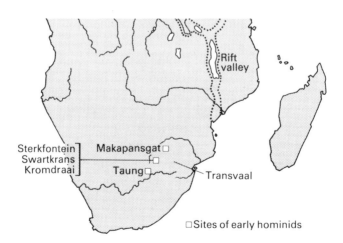

Fig 3.6   The earliest finds of *Australopithecus* came from the Transvaal in South Africa.

Fig 3.7   Olduvai Gorge is a remarkable landform as well as a fossil gold mine. In
this photograph, Louis Leakey (centre) shows visitors some features of
interest. (*Des Bartlett, Courtesy National Geographic Society*)

Richard Leakey and associates near Lake Turkana and of Drs Mary
and Louis Leakey at Olduvai Gorge and their meticulous recovery of
data, have confirmed many of the claims of Dart and his colleagues.
The Olduvai localities have yielded superbly preserved undisturbed
'living floors'—that is, occupation sites where the hominids made their
artefacts, butchered animals, and conducted the day-to-day business of
living. The preservation of such archaeological data for thousands of
years is rare; the fact that the Olduvai occurrences have been dated
with confidence to almost two million years ago makes the remarkable
preservation of these sites all the more astonishing.

Olduvai Gorge is a deep valley that cuts through the Serengeti Plain
of Tanzania; the main branch of the gorge is about twenty miles long,
and the deposits of Pleistocene and Pliocene age reach up to 100 metres
in depth (Fig 3.7). Today there is a temporary stream that flows through
the gorge, but during the earliest phase of the Pleistocene there was a
sizeable lake whose shores served as camping sites for the hominids.
This lake was formed by the action of volcanoes about two million
years ago, volcanoes that laid down huge lava flows which served as a
barrier for some local streams.

The earliest period from which we have deposits in this area was one
of rapid deposition of river-laid, air-borne, and lacustrine materials that
sealed in the hominid occupation sites. It is this very rapid deposition

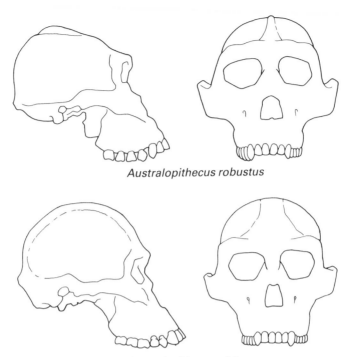

Australopithecus robustus

Australopithecus africanus

Fig 3.8 The skulls of the robust (above) and gracile *Australopithecus* (below) are here compared. Note the larger braincase but narrower cheekbones of the gracile form.

that accounts for the kind of preservation that characterizes the archaeological sites of the late Pliocene and earliest Pleistocene at Olduvai. Geological evidence suggests that during the period in question the climatic regime was not very different from conditions obtaining today—a semi-arid climate with a highly seasonal rainfall pattern.

Once the earliest hominids had made the transition to woodland and savanna, this rich environment must have offered a vast amount of resources: edible roots and tubers, fruits from the trees that dot the plain, large numbers of small amphibians and mammals, as well as the larger animals that came to the shore of the lake to drink. Only the big cats and other carnivores like the hyena, which hunt and scavenge, were capable of exploiting the food resources offered by the huge mammalian herds; it was a niche that the early hominids came to occupy with success.

The hominids themselves appear to be quite varied in form, the natural result of the rapid radiation of an ancestral species into the extensive savanna habitat. Most human paleontologists agree today that there are at least two clearly distinguishable forms: a small gracile form and a much larger more robust one (Fig 3.8). If these represent

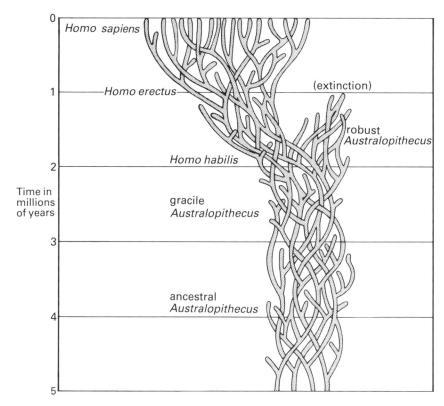

Fig 3.9   This evolutionary 'tree' diagram of the hominid lineage is designed to emphasize the complexity of the evolutionary process. Each evolving species comprises unnumbered varieties and populations. The diagram is intended to indicate the succession and relationship of the various hominid forms represented in the fossil record.

more than one species in the same habitat—and the two forms do occur together on several sites—then Gause's 'law' states that they would have had to occupy different ecological niches. Several ingenious hypotheses have been offered, suggesting that the large and small forms did indeed exploit different food sources and that these patterns of food-getting imply significant differences in behaviour. Today we believe from the quite extensive fossil evidence that the small, gracile form probably gave rise to *Homo* (and eventually *Homo sapiens*) while the large, robust form became extinct about one million years ago. Because they have so much in common, we believe they shared a common ancestor between 3 and 4 million years ago (Fig 3.9).

The robust form is characterized by very large molar teeth and quite small incisor and canine teeth—a dentition clearly evolved to chew

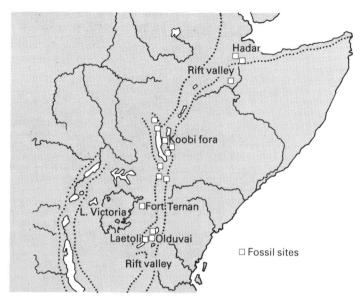

Fig 3.10   Large-scale map of East Africa showing some important fossil-bearing
sites. Almost all occur along the great Rift Valley (dotted)

large quantities of tough vegetable foods and unsuited for the consumption of meat. It is widely believed that the robust australopithecines had evolved a fully vegetarian diet derived from the savannas of
the areas they occupied (Fig 3.10). Savanna herbivores are generally
larger than their carnivorous predators, and often achieve large size
(e.g. elephants, hippopotamus, etc.) so we can predict that a vegetarian
form would show rapid increase in size. This is born out by the fossil
evidence. These forms are perhaps 1·5 m high and weighed about 70 kg.
Their cranial capacity (an indication of brain size) is 500–530 cc, and
they were bipedal, though they probably did not walk as well as we do.

The smaller gracile form (Fig 3.11) had a different dentition.
Although the molars show a slight increase in size relative to the front
teeth, this is not greatly marked, and the incisors and canines are much
less reduced. In fact their teeth are not very different to those found in
modern man. Their size is small, the early forms being no more than
just over 1 metre in height, and weighing perhaps 36 kg only. The
retention of effective incisors and canines, though small compared with
those of the apes and monkeys, does suggest a more carnivorous lifestyle
in comparison with that of the heavy robust populations. This is amply
borne out by the later archaeological evidence. Their cranial capacity
varies from about 400 cc among the earliest forms (little bigger than

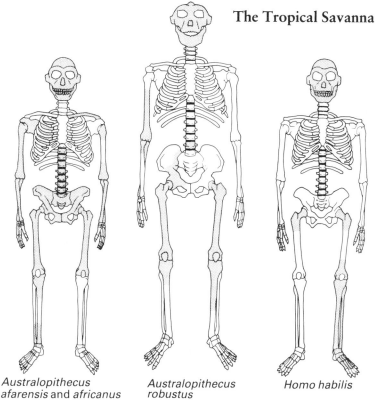

Australopithecus
afarensis and africanus

Australopithecus
robustus

Homo habilis

Fig 3.11   These three skeletons are shaded to represent the extent of fossil hominid
material available for study. On the left the gracile *Australopithecus*, in the
middle the robust form, and to the right, the earliest *Homo*, *Homo habilis*.

that of a chimpanzee) to over 500 cc in the later forms. As early as 3·5
million years ago, these creatures appear to have been fully bipedal.
Probably the most significant deduction drawn from a foot skeleton
found at Olduvai is that the infant hominid could not cling to his
mother as other higher primates do but needed to be carried. Some of
the implications of this for adult female behaviour and social organization are discussed below.

HOMO HABILIS: OCCUPATION SITES

The earliest occupation site excavated at Olduvai Gorge directly over-
lies a basalt flow, and it has been dated by radioactive techniques to
almost 1·9 million years ago. This site is an important one on several
counts. It has yielded remains of the first archaeologically documented
structure—an oval about 5 m long, consisting of small piles of basalt
rocks (Fig 3.12)[1]. Were such a stone ring, and patterning of stone and
bone within the ring, to be found on a Late Pleistocene site, it would

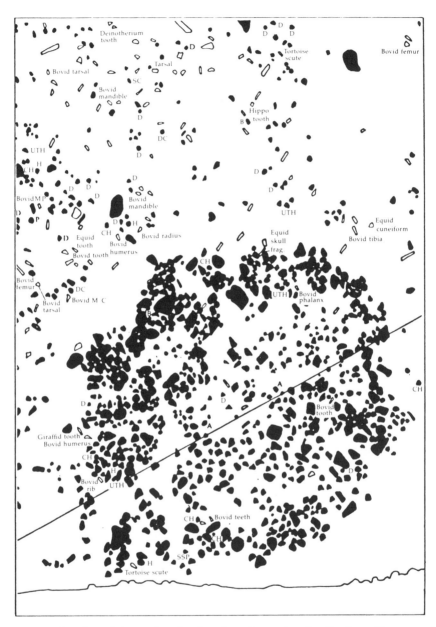

Fig 3.12   The plan of the ancient land surface shows the distribution of bones and stones at a site in Bed I of Olduvai Gorge discovered by Louis and Mary Leakey. The circle of rock suggests the base of a windbreak or simple hut-like structure, the earliest known. (From Leakey[1])

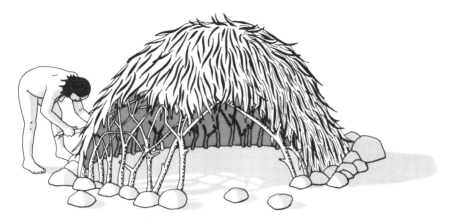

Fig 3.13   The shelter found at Olduvai was probably not very different to the kind of hut still built by many African peoples. This drawing shows a !kung/Bushman hut of a kind built at the present time.

certainly be interpreted as the foundation for a hut or tent; while its specific function is not now known, it must have been at least a hunting-blind or windbreak, and it might very well represent the foundation for a dwelling (Fig 3.13). The area to the north and west of the stone ring is strewn with stone implements and animal bones. The most numerous stone artefacts are unmodified flakes—small, thin pieces of stone with excellent cutting edges. Such unmodified flakes make up almost three-quarters of the artefacts at this locality. There are also numerous finished tools of the kind useful for heavy duty bashing and chopping, in addition to smaller numbers of tools useful for piercing, cutting, and incising (Fig 3.14). Associated with the tools is a large amount of tool-making debris, mostly of quartz, which does not occur locally and must have been carried in. The animal bones range from remains of small forms, such as tortoise, to a large extinct elephant (*Deinotherium*) as well as bones of many bovids and equids. Giraffid, hippopotamus, and wild pig are also present though in smaller numbers. The presence of crocodiles as well as remains of a plant resembling papyrus are taken as indications that this occupation was a lakeside one (Fig 3.15).

Another well-preserved site at Olduvai occurs about midway through the lowest bed of deposits and has been dated to about 1·75 million years ago. This was the first of the undisturbed archaeological sites from this time range discovered by the Leakeys, and the excavation was an extensive one that uncovered more than 280 m². The major portion

Fig 3.14   Olduvan tools from the lowest beds constitute the earliest type of stone industry that we know. The simplest tools are of two kinds: a rounded cobble from which flakes have been struck to make a cutting edge (a core tool) and the sharp flakes themselves (each a flake tool).

Fig 3.15   The ancient fauna represented at Olduvai was even richer than that of the Serengeti plains today and included giant forms. This reconstruction gives some idea of the variety of extinct animal life together with an early giraffe.

of the archaeological remains occur within a dense concentration covering about 28 m², with lesser concentrations and scatters of stone and bone in the remaining area. The tool types are very like those in the earlier site, but there is a higher percentage of purposefully modified scrapers and reduced frequency of choppers. The most common artefact, as in the earlier site, is the unmodified flake (almost 90% of the total inventory). The implements are associated with bits and pieces of animal limb bone in the central concentration; the peripheral areas have yielded animal jaws, pelves, and scapulae along with natural stones and larger artefacts. Most of the animal remains are of small mammals, fish, reptiles, and birds; however, there are also an impressive number and variety of large forms: bovid, equid, giraffid, and wild pig.

Toward the top of the deposits laid down in the lowest bed are the remains of an extinct form of elephant. The skeleton was partially articulated and was associated with stone implements. Most of the modified artefacts are of the heavy-duty sort, but here too the most frequently occurring type is the unmodified flake. There are other bits of large mammals at this location, including bovid, equid, and wild pig.

These three sites at Olduvai Gorge establish firmly that at least some of these hominids were tool-makers, and the manufacturers of a much more varied inventory than had been supposed. In attempting to reconstruct the technological capabilities of these early hominids it must be remembered that the only data we have to work with are those items that have been preserved: items of stone, a small percentage of preserved bone, and very rarely wood. Such differential preservation removes from our scrutiny an undoubtedly large number of items of perishable materials. Even under the remarkable conditions of preservation at Olduvai Gorge, objects made of grasses, rushes, wood, hide, or other organic material would not appear on ancient living-floors.

The presence of many large mammals at the three locations in association with stone implements, especially with high frequencies of flakes with cutting edges, indicates unequivocally that the hominids were butchering and consuming game. Whether they were actually killing the animals or scavenging the kills of other predators cannot be determined on the basis of the present evidence, but the absence of spear points or wooden spears does not necessarily imply scavenging rather than hunting. There are several groups of living hunters whose successful hunting depends more on cunning, swiftness, and knowledge of their prey's habits than on the possession of sophisticated weaponry. In the game preserves of modern Africa, where the use of guns is prohibited, animals can be approached relatively easily. It appears that shyness of humans is learned behaviour on the part of most herbivores.

The significance of these and other sites at Olduvai is that *Homo habilis*, as early as one and three quarter million years ago, had already become systematic consumers on at least two trophic levels. It is not, however, intended to imply that the hominids consumed mostly meat; the diet of hunter-gatherers in comparable environments today is made up principally of vegetable matter. The fact that these hominids possessed large molars showing intensive wear and chipping, testifies to their heavy dependence on coarse and fibrous plants. However, the implication of the presence of butchered animals on all the Olduvai sites is that meat was a regular part of their diet, and these early hominids were surely well on their way toward becoming successful social predators together with the lion, hyena, and wild dogs. The systematic production of cutting implements means also that they were freed from the constraint of the general rule that predators tend to be larger than their prey. This rule on predator size, however, has been broken a number of times, especially by those species which have adapted to life on the savanna. It is clear that predator species hunting in packs or prides can kill animals much larger than themselves and obtain enough meat at one kill for a whole social group. African wild dogs demonstrate this observation most strikingly: working together as a pack of 10-20 animals, these dogs can bring down zebras that weigh up to 250 kg though the dogs themselves weigh only about 18 kg each. Thus cooperation in the hunt opens up an enormously greater variety and quantity of prey to a social predator. In view of the high density of herbivores on the savanna, which may be expressed as 30-50 kg body weight per hectare, the potential supply of meat is obviously immense. The overall ratio of numbers between prey animals and predators on the savanna is believed to be about 100 : 1. (This is not determined by prudent management on the part of the predators, but by a number of variables including their ability to kill versus the ability of the prey to escape.) These two components of the biomass tend to remain roughly in equilibrium.

The behavioural implications of the archaeological remains from Olduvai are far-reaching. Cooperation between individuals in obtaining meat would have been essential, and the butchering of large 'packages' of meat such as elephants suggests further that food must have been shared among individuals. Except in a very limited way, such food-sharing is not found among the other higher primates. Among all known hunter-gatherers obtaining meat is a male task, while the collection and preparation of plant food is done by women and children; such an arrangement is obviously based on the sharing of different food resources by different members of a social unit.

While we have no direct way of knowing the size or nature of the hominid social unit, we can make some deductions based not only on the kind of game butchered but also on the rather improbable-sounding basis of the *Homo habilis* foot. The foot bones recovered from Olduvai are those of an organ already clearly adapted to bipedal locomotion: the big toe is in line with the others and is not an opposable, grasping digit as in monkeys and apes. Among non-human higher primates the infant clings with his fore and hind feet to his mother's fur, leaving her free to move about with other members of the group while offering food and protection to her offspring. In the absence of a prehensile toe, the hominid infant is utterly dependent on its mother for holding during nursing, and for being carried about, so that we can be sure that the nature of infant care had changed drastically. Not only was the infant more dependent on its mother, but the mother was therefore less mobile and more dependent for food and protection upon other adults. This might well have been the context in which a male assumed responsibility for the feeding and protection of a female (or females) and her young, and the selective advantages of such an arrangement for the successful rearing of the young are evident. The loose linking together of these groups into bands would have offered the further advantages of supplying a regular labour force of cooperating males for hunting or scavenging and of both sexes for the butchering and consumption of meat.

It is perhaps worth adding that there is a real sense in which a predator's life contains leisure, which is hardly present in a herbivore's existence. Most herbivorous animals spend their entire waking life searching for food, eating it, or (in the case of bovids) chewing the cud. For most large predators, it is usually sufficient to hunt every other day, or even every third day. This means that in East Africa, dogs or lions can be seen basking in the sun, lying on rocks gazing at the country, or playing with their young. If the hominids were already able to partake of such leisure, they would surely have been in a better position for the evolution of exploration, experimentation, and social communication. Leisure is the privilege of the predator.

Given the enormity of today's human population numbers and densities, it is difficult to conceive of even our very early ancestors as a numerically insignificant component of the world's animal population. Humankind's current status as the prime predator and disturber of the world's ecosystems tends to obliterate the fact that for over four million years hominids were probably very few in number. Ecologist Edward Deevey has guessed that at about two million years ago there was a total hominid population of 125 000 on the African continent and we may guess that their mean density on the savanna was not so different from that of lions today at 0·05–0·25 per sq km (0·1–0·5 per square mile). The rate of technological innovation was exceedingly slow during this period (approximately 5 to 1 million years ago), and *Australopithecus* and *Homo habilis* were participants in the longest-run equilibrium system of any hominid. Their capacity to alter the ecological equilibrium significantly, let alone exhaust local resources, was limited by their technological capabilities as well as by their small population numbers. Certainly they were incapable of producing the major ecological catastrophes that modern humans like to term 'the conquest of nature'.

THE HADZA

The group referred to here as the Hadza are the eastern population of hunters and gatherers living near Lake Eyasi in Tanzania, not far from Olduvai (Figs 3.16–3.18). The Hadza live in a relatively dry savanna

Fig 3.16 Tlanjalaga, a Hadza, prepares an arrow shaft. His hunting bow and arrows lie on a rock beside him. The children learn adult skills by watching closely as well as in play. (*James Woodburn*)

Fig 3.17   A Hadza woman divides up strips of zebra meat, which are hanging on a branch beside her grass hut. (*James Woodburn*)

Fig 3.18   Hadza women use their digging sticks to extract a wide variety of edible roots, many of which occur in dry stony areas. The roots are carried back to camp in the woman's leather carrying garment which is slung over her shoulder. (*James Woodburn*)

area to the east of the lake—grassland interspersed with areas of scrub bush and acacia trees. About 400 of these hunter-gatherers occupy over 2500 km² at an average density of 0·4 persons per 2·5 km² (square mile). Game abounds in Hadza territory: several species of antelope, together with giraffe, rhinoceros, baboon, warthog, and elephant which are the common large animals. There is a rich small-species fauna too, including hare, tortoise, and hyrax. All of these forms are hunted by the Hadza with the exception of the elephant. Tuberous roots and berries comprise the staples of Hadza diet, supplemented by honey and the grubs of wild bees. Though water is scarce in the dry season, there are many sources, well-distributed, during the rainy season.

The Hadza diet consists of perhaps as much as 80% by weight of plant material with the remainder made up of meat and honey. James Woodburn, the ethnographer who studied the Hadza, estimates that an average of only two hours per person per day is spent in subsistence activities[2]. Predation has brought them leisure. No plants are cultivated nor is there any effort toward systematic cropping or conservation of animals[3]. Men and women both go daily to obtain vegetable food and in this activity are largely independent of each other. Hunting is done by males with bow and poisoned arrows, and it is essentially an individual activity or one involving only a few males. Small game are cooked and consumed on the spot by the men, but larger animals are brought back to camp and their meat is rapidly distributed[4]. One other important source of meat for the Hadza is scavenging, which is done by careful observation of the movements of vultures. By this means the Hadza can locate not only kills of predators, whom they do not hesitate to drive away from the carcass, but also animals which have died naturally, and which provide an easy source of meat. The Hadza usually practise no form of food preservation or storage; the techniques are known but are considered a waste of time and effort.

Hadza camps are situated within areas of rocks and/or trees and consist of simply constructed brush huts. The Hadza live in dispersed camps whose average size is about eighteen adults, but settlement size varies with the season: in the wet season, when water resources are widely distributed, camps are small. The larger aggregations occur during the dry season when water resources are less well distributed and when the larger animals are hunted. Woodburn thinks that size of game is the critical factor in accounting for the larger Hadza groupings. When a large animal has been killed, people know there will be plenty of meat to go around, and meat is a highly prized food.

Among the Hadza there is a total absence of any rights over the land or its resources. They do not attempt to exclude outsiders from their territory, nor do they make individual claims to any portions of the land. Any Hadza is free to collect plants or hunt anywhere he wishes. They maintain only minimal trade relations with neighbouring groups and constitute an essentially independent socio-economic system; in this they are different from the Mbuti pygmies.

Given their fairly casual attitude toward subsistence, the Hadza nutritional level is remarkably high, and their children are in general free of the parasites and diseases associated with the commonly seen tropical malnutrition. Their environment provides much more food than they can consume, and they have been known to take in neighbouring agricultural people whose crops are subject to recurrent failure. The amount and diversity of food in their area, together with low population density, make the Hadza way of life a secure one. Today, we tend to think of these 'primitive' hunters as very poor people, eking out a bare subsistence from food they have great difficulty in obtaining. This picture is quite wrong for the Hadza, nor is it apparently true for a great many other hunter-gatherer groups, even those living in less congenial environments. The success of these technologically unsophisticated people appears to be related to the essential fact that they live well below the carrying capacity of their environment, and this in turn is directly dependent upon small population numbers and low densities.

In the case of nomadic hunter-gatherers generally, population is always subject to both biological and strict human behavioural controls. Widely spaced births are maintained by their natural practice of nursing children at the breast up to 3 or 4 years of age. An additional brake to population growth is achieved by infanticide, abortion, and taboos against intercourse for nursing mothers, all of which are very common practices. By these means they are able to replace those members of the society who have died without significantly increasing the numbers of births over deaths. Population control among humans, as in many animal species, is one of the critical factors in maintaining equilibrium in an ecosystem (see Chapter 11).

SUMMARY AND CONCLUSIONS

We have seen that the tropical savanna offers hominids an excellent environment for hunting and gathering. The plant resources, while less varied than those of the tropical rain forest, are a reliable and predictable source of food. The large herds of herbivores provide meat in large quantities, which can be obtained both by relatively simple hunting techniques and by scavenging. We have examined the evidence of two adaptations to this biome: one that goes back into the early time ranges of prehistory and the other a group of modern hunter-gatherers who extract a well balanced diet from wild foods with a minimal amount of energy expended. Both of these examples are fairly stable and long-lasting equilibrium systems, despite the obvious differences in the hominids themselves and in their technological capabilities. What these groups share is a pattern of environmental exploitation that disrupts the existing relationships between other systemic components far less than pastoralism or agriculture. Neither the early hominids nor the Hadza show evidence of predatory behaviour toward plants or other animals that would seriously affect future food supplies in the environment though fires may spread by chance during the dry season. In the case of the early hominid population, density was probably also low, and maintained for a long period of time without the use of birth control mechanisms. Infant mortality was probably quite high, and certainly *Homo habilis* and their predecessors, who did not yet have the controlled use of fire, were probably a food source themselves for carnivores. The Hadza, like so many other nomadic hunter-gatherer groups, undoubtedly limit population growth by cultural means, thus insuring themselves against both exhausting local resources and human crowding.

The agricultural and pastoral neighbours of the Hadza have steadily encroached on their lands and reduced the area available to them. Finally, in 1964 and 1965, most of the Hadza were settled by the Tanzanian government with the intention of teaching them to live by peasant farming.

References

1   Leakey, M. D. 1971 *Olduvai Gorge: Vol. 3 Excavations in Beds I and II, 1960–1963* (Cambridge University Press).
2   Woodburn, J. 1968 An introduction to Hadza ecology, *and* Stability and flexibility in Hadza residential groupings. In *Man the Hunter*, R. B. Lee and I. DeVore, eds. (Chicago: Aldine Publishing Co.).
3   Woodburn, J. C. 1980 Hunters and gatherers today and reconstruction of the past. In *Soviet and Western Anthropology*, E. Gellner, ed. (London: Duckworth).
4   Woodburn, J. C. and S. Hudson 1966 *The Hazda: the food quest of an East African hunting and gathering tribe* (16 mm film).

4 The Temperate Forest

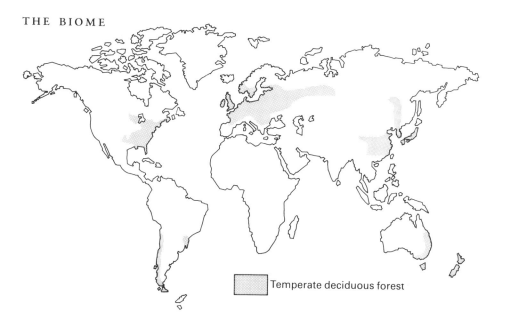

Temperate deciduous forest

Fig 4.1 The temperate forest is familiar to most English-speaking readers that live in the northern hemisphere, and was the aboriginal environment of most of Europe and the Northeastern USA. The vast majority of European temperate forests have been cleared for agriculture since the soil and climatic conditions are ideal for intensive food production without irrigation. If further forest clearance were to occur, wildlife diversity would be finally and drastically reduced and hardwood production would be terminated.

The temperate woodland biome is distinguished sharply from the two biomes which we have considered so far in that it is found sufficiently far north to be subject to seasonal fluctuations in temperature. There are usually two or three months of frosty weather in winter, though the degree and period of cold varies. Trees are typically deciduous, and this striking change in the apppearance of the forest, together with the strictly seasonal flowering and fruiting, emphasizes the seasonal variation in temperature. The commonest trees of the forest are the maple,

Fig 4.2   A mature English oakwood is a lovely example of the kind of climax forest
which will develop in the North Temperate zone when climatic conditions are
appropriate. There may be considerable ground cover in the spring before the
leaves develop, but during the summer there is very little. The diversity of species
is much lower than is found in the tropical rain forests. (*Natural History
Photographic Agency*)

beech, oak, hazel, and elm. Today this biome is widespread throughout
Europe, the eastern USA and parts of Asia, especially China and Japan
(Figs 4.1, 4.2). It is also the biome that has suffered the most extensive
destruction for timber or through clearance for agriculture and grazing
in Europe and North America. In this case the natural climax of
woodland is no longer present, but the biome is maintained artificially
at a different level of stability, often described as a disturbance climax,
or disclimax. The temperate forest biome can be stabilized more readily
at a disclimax than can the tropical biomes when they are altered by
human activity (see Chapter 8). The stability of this biome we owe
primarily to its diversity; it carries a range of species second only to the
tropical rain forest. It has good powers of recovery from human damage
for this reason and because of the rainfall pattern and quality of the
soil associated with it. Our so-called Western civilization has achieved
its greatest development in this biome.

The presence of forest rather than grassland in this region is due to a fairly evenly distributed rainfall (750-1500 mm), humid summers, moderate temperatures, and few droughts. Lakes are common, with permanent streams and rivers. The presence of deciduous trees in turn means that for part of the year sunlight reaches the ground, and as a result there are well developed layers of herbs and shrubs and a much richer soil life than is found in the tropical rain forest. At the same time the rate of deposition of foliage litter is greater than in the tropics and the litter itself contains more abundant mineral salts and organic matter. This and the rainfall pattern result in less leaching of minerals than we find in the tropics, and the constant addition of organic material to the topsoil maintains its fertility. As a result these *podsolic* topsoils are fertile when cleared for agriculture, and more stable than the tropical soils, though without care they do not stand up indefinitely to crop production.

The varied flora and fauna make the forest a suitable environment for hunter-gatherers; it is well endowed with fruit and nuts, and with a wide variety of small game, such as birds and rodents, as well as the larger deer, together with pigs and bears, among other species (Fig 4.3). The seasonality of the food resources, whether herbage, fruit, or nuts, has resulted in seasonal movement of herbivores and their attendant carnivores. In particular, the deer tend to gather into larger aggregations in late summer and autumn and disperse singly or in pairs in winter or in spring. These movements are related to the concentration of food supplies available during the nut season, and the need for dispersal to find food at other times. These factors imply seasonal hunting and gathering for humans, with a seasonally determined settlement pattern.

The temperate deciduous forest is very much more productive than the savanna, but a much smaller proportion of this productivity is available for grazing and browsing herbivores, because grass and low shrubs are sparse.

Human adaptations to temperate regions require more than techniques of hunting and gathering. Having evolved in the tropics, we are adapted to survive within a range of temperatures around 27°C (80°F). This particular temperature is called the *critical* temperature and it is that at which humans neither cool nor warm their bodies; neither sweat nor shiver. Although we can withstand surprisingly large variations about this temperature, we know that for comfort and survival we usually maintain an artificial microclimate around ourselves as near as

Fig 4.3   The red deer is a typical forest creature which at one time was very abundant in Europe. The ancient forest also carried wild boar. (*Natural History Photographic Agency*)

possible to the critical temperature. It seems probable therefore that when our ancestors occupied the temperate regions of the northern hemisphere they had at their disposal either effective shelter, clothing (skins), or fire, and probably all three of these. The evidence of some human adaptations to the temperate zones can be clearly seen in the archaeological record. In this chapter we shall examine an ancient settlement of hunters and gatherers in China, and the recent (but now extinct) adaptation of hunters, gatherers, and agriculturalists in North America.

THE CAVE AT CHOUKOUTIEN

There is considerable uncertainty about the time at which early humans from Africa (*Homo habilis*) originally entered and spread throughout Europe and Asia, but we do know for certain that their successors were widely dispersed by about one million years ago, if not earlier. By that time we have evidence of humans in southeast Asia and southern Europe. By about 500 000 years BP the archaeological record is more

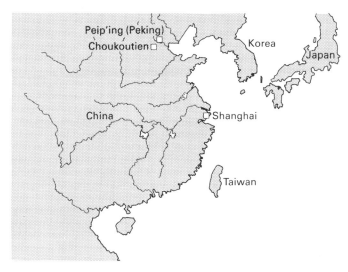

Fig 4.4 Choukoutien lies close to Peking and is only one of numerous fossil-bearing sites in China. There are exciting possibilities for further research in China into early human adaptations to temperate conditions.

complete and is better preserved, as humans expanded their range northward from sub-tropical to temperate regions. One of the most important sites known from this period is that at Choukoutien, a village near Peking in central China (Fig 4.4). This site lies in the temperate woodland biome today, and we have evidence that at the time this site was first inhabited by people, the climate and biome were quite similar to those of the present day. In view of the extent and importance of the archaeological investigations carried out at this site, we shall consider it in some detail.

The village of Choukoutien lies at the foot of a range of hills, some 36 m above a wide fertile plain, The site itself comprises the major part of a small hill near the village. The side of the hill consists of the consolidated filling of an ancient cave of immense size, which was at least as much as 50 m in height, and 175 m in length. In the course of time, the roof of the cave, which was formed out of the limestone bedrock, has been eroded away, so that the filling is now at the surface of the ground.

The word Choukoutien means 'dragon bone hill', and it was long known by the local Chinese as a place where fossil bones could be gathered. It was first recognized as an important palaeontological site in 1918 by the Swedish palaeontologist J. G. Andersson. Excavations soon after revealed three human teeth together with quartz artefacts

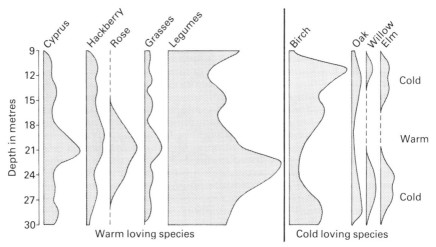

Fig 4.5   The analysis of pollen from an excavation can be very revealing about the environmental conditions at a particular site and time. The kind of results that can be obtained by this painstaking research can be seen in this diagram summarizing a Chinese study made at Choukoutien. The flora indicates that during the period of deposition in the cave, the climate changed from cool to warm to cool—which represents a warm period in the Pleistocene nearly 500000 years ago. (After Hsu[3])

and a mass of fossilized animal bone. From 1927 to 1937 extensive excavation took place and thousands of tons of the consolidated cave filling were removed by drilling and blasting. All the faunal and cultural remains were recorded, but the excavation was made in blocks of one cubic metre. In these early days there was no horizontal recording of living floors as at Olduvai (Chapter 3) although such living floors undoubtedly existed.

In spite of the techniques which were originally employed at Choukoutien (which compare unfavourably with today's methods), a large amount of information is available through the pre-war Chinese publications[1-2]. Recently new research has been undertaken by the Chinese at this site which has proved of great value. The following data are derived from both the prewar and the more recent Chinese publications.

Chinese botanists have recently collected pollen samples from much of the cave filling and analysed the flora which existed in the vicinity of the cave at the time of deposition[3]. Throughout the pollen profile, herbs, shrubs, and deciduous trees are equally common, and there is a small proportion of coniferous pollen (about 10%) (Fig 4.5). The samples, however, were carefully taken throughout the entire depth of the cave filling, and an examination of these different samples in terms of genera of plants reveals that the base and top of the deposit contain plants associated with a cold climate, while the main, central part of the deposit contains woodland plants associated with a warm temperate climate. Towards the centre of the deposit there is the zone of maximum warmth containing pollen of plants found today somewhat south of Peking. This suggests that at that time the climate at Choukoutien was somewhat warmer than at present. This sequence of cold–warm–cold in the flora of the cave deposits indicates that they were laid down during an interglacial period: that is, between two of the ice ages or cold periods which are known to have occurred in the northern hemisphere during the last million years. We do not know, however, the length of time represented by this warm period, although it certainly fell between 600 000 and 300 000 years BP. We also know for sure that the warm spell, which covers the period of occupation of the cave, was characterized by temperate deciduous forest, with its characteristic oak, beech, and birch, together with a wide variety of shrubs, herbs, and grasses.

The importance of the site at Choukoutien is due primarily to the large number of bones of early humans which were recovered from the cave deposits. These fragmentary remains belong to the species of *Homo* designated *Homo erectus* which preceded modern humans (*Homo sapiens*), and succeeded *Homo habilis*. The remains represent some 40 individuals, men, women, and children, including parts of 14 skulls, some facial bones, jaws, teeth, and a few limb bones. The remains occur mainly towards the middle of the deposit, when the climate was warm, though a few fragments date from colder times.

Although the remains are fragmentary, they tell us a good deal about the occupants of the cave. In fact, these bones constitute one of the largest samples of prehistoric populations that we possess. A study of them indicates that these people were not strikingly different from ourselves except in the region of the head (Fig 4.6)[4]. Their brains were smaller than ours, but a lot bigger than those of *Australopithecus* or *Homo habilis*. The mean cranial capacity of 1050 cc falls just within the range of the brain size of modern humans (1000 to 2000 cc). In comparing *Homo erectus* with *Homo habilis*, we see not only a general increase in body size but almost a doubling of mean cranial capacity over a relatively short span of time, evolutionarily speaking. Many authorities on human evolution are convinced that this rapid growth of the brain is the result of selective pressures created primarily by a social, hunting way of life. Increased efficiency in the hunt, increased capacity

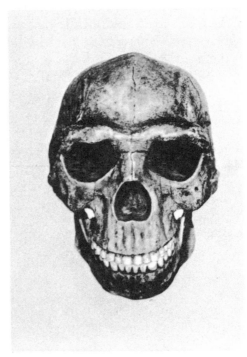

Fig 4.6   The fossil human remains from Choukoutien have been reconstructed to give us a good idea of the form of *Homo erectus* at this time. This reconstructed skull has heavy brow ridges and jaws and a brain of about 1100 cc. It measures about 10 cm across. (*British Museum, Natural History*)

to perceive and process information, and increasing technological proficiency are components that operated in a system of positive feedback whose end result was the transformation of the slight, small-brained *Australopithecus* into *Homo habilis* and eventually *Homo erectus*, a hunter of considerable skill.

The masticatory apparatus (the teeth, jaws, and associated musculature), on the other hand, was still heavily built, and this suggests that their diet, like that of *Australopithecus* and *Homo habilis*, required extensive chewing. Apart from these differences, the people of Choukoutien were almost modern in their anatomy, and ran and walked with proficiency. Their hunting adaptations also undoubtedly involved flexible and complex behaviour which may have been associated with the beginnings of language.

A review of the faunal remains from Choukoutien is of considerable significance in our interpretations of human behavioural adaptations

Fig 4.7 Bears probably competed with people for the shelter of caves during a period of hundreds of thousands of years. Much later, their skulls were treated as sacred objects by Neandertals, perhaps because of the giant size of the bears and the danger they constituted. They were, however, probably fairly easy to kill if sought out while they were hibernating in the recesses of the caves.

to the woodland environment. Unfortunately, we do not have very precise records of the fauna associated with the human remains, but we have enough information to enable us to make some useful deductions.

In the first place, only a small number of animals besides *Homo erectus* would naturally be likely to occupy the cave and die there. These include the two giant carnivores, the brown bear and the short-faced hyena, which might have carried their prey into the cave (Fig 4.7). Almost all the remains of herbivores, and most of the other carnivores found in the deposits were brought in by humans.

The excavators at Choukoutien distinguished three main cultural zones—not just living floors, but phases of the deposit which were rich in stone artefacts, bones, and ash. These zones represent the period of most intensive human occupation: between the zones, we find remains of brown bear, which may well have occupied and hibernated in the cave when humans were not in occupation.

Since the majority of the faunal remains come from the cultural zones, we can consider that the fauna as a whole represents the animal food resources of these cave dwellers. Among the herbivores the commonest are the red deer, three other species of deer, sheep, zebra, pigs, buffalo, and rhinoceros. We also find traces of macaques, bison, and elephant. What is striking about this very diverse fauna is that the deer represent about 70% of the total. Among the carnivores identified, we find the commonest are wolves, the fox, badger, leopard, and various other cats. The evidence suggests that at least some of these must have been the food remains of the inhabitants: the total number of species is in the region of forty-five, and this represents a very diverse larder.

Choukoutien is also remarkable for the preservation, in the cultural zones, of vegetable food remains. In the middle zone (B) there is a layer several inches thick of shells removed from hackberry seeds[5]. The seeds of hackberry (*Celtis*) are eaten today by Indians in the southwestern USA, and there is no doubt that this deposit represents human food remains. The Choukoutien pollen diagram includes a number of trees and shrubs which bear edible fruit and seeds, and there is little doubt that these, together with green vegetables, formed an important resource for the cave dwellers. This is one of the few prehistoric sites where preserved vegetable food remains have been recovered.

Artefacts are found throughout the three cultural zones and are on the whole of similar type. Thousands of broken stones were found, but only a few hundred implements, with a fairly large range of types: besides cores and flakes, there are choppers, scrapers of various degrees of sophistication, and pointed implements (Fig 4.8). These are mostly of quartz, but occasionally of limestone or chert (flint). There is some slight advance in the preparation of raw material in the upper zones, but this is not very striking considering the time interval that may be represented here. There are scratches and incisions on many of the bones which occur in the deposits, which indicate the use of flakes and scrapers, and longbones cracked for marrow indicate the use of heavier rocks. It has been claimed that some of the longbone fragments were made into tools, and this seems likely from what we know of early technology elsewhere[6].

One of the most fascinating features of the Choukoutien cave deposits is the extensive deposition of charcoal, ash, burnt wood, and burnt bone. These ashy deposits are up to seven metres thick in places

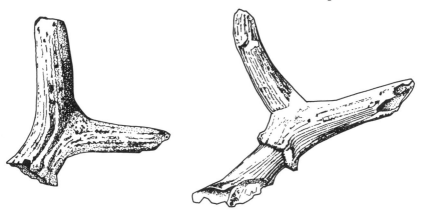

Fig 4.8 Antler tools with worn tips, as well as simple stone tools, were found in abundance at Choukoutien. They were probably used as picks and digging tools. The game animals were by now supplying not only food but essential tools and manufacturing materials for needles, thongs, clothing, and much else.

and this suggests that fire may have been kept burning continuously in the cave. The presence of many burnt bones implies that meat was probably cooked; and fires would not only serve for cooking and warmth, but for protection against predators such as the sabre-toothed tiger.

BEHAVIOURAL IMPLICATIONS

From this evidence we can piece together something of the life of *Homo erectus* in the temperate woodlands. In the first place it is clear that they lived for at least part of the year in caves, and that with wood for fuel they relied extensively on fire for warmth and protection. While we deduce that they lived extensively on fruit, mushrooms, and other vegetable foods, we have empirical evidence that they consumed large quantities of nuts and were effective hunters. Though deer hunting was their main means of getting meat, they also consumed a wide variety of other species, and among them a number of carnivores. They not only had a very diverse diet, therefore, but were exploiting three trophic levels. We can conclude that their food resources were secure and plentiful.

As a living site, this particular cave was well chosen. The cave mouth had a view across the river valley, over the trees and clearings through which the game passed on its way to get water. A variety of materials for tool making were at hand: besides stone, there was wood and bone.

The stone choppers, scrapers, and sharp flakes were no doubt used to work wood and bone as well as to cut up meat. Other animal products would also have been prepared: skin, bone tools, and perhaps leather thongs and cords from ligaments, sinews, and gut.

We can make a few extrapolations about the social life of these people. We know that by this time humans had evolved a brain not much smaller than that of today. It follows that babies would have had much larger heads than previously, and were therefore born at an earlier stage in their development to allow safe passage through the birth canal of the mother. This evolutionary change increased the importance of maternal care in the early years: babies would have needed constant attention from their mothers for at least three years, if not more. We can suppose therefore that mothers with young children would not have been very mobile and would have remained in the area of the cave, looking after their children, preparing food, and tending the fire. Older children, and women not carrying babies, would have been mainly responsible for the collection of vegetable foods, and for carrying water from the river in skins.

The data from Choukoutien suggest a pattern of life centred on a home base. Even if the cave was not occupied the whole year, the maintenance of the hearth, and the preparation of tools and skins would have taken place in the area of the cave and the cave mouth. The importance of fire would have been great, and if they had only learnt to maintain it rather than make it (having captured wild fire), its maintenance would have become a matter of great social importance. The possibility of cooking food would increase their extractive efficiency by opening up more food resources: vegetables indigestible in a raw state. The warmth would have been essential in winter in this climate, and the fire itself could be used to harden spears.

The complexity of the social life and technology of these people suggests that the evolution of language was possibly quite far advanced. The separation of the band into groups of males and females, the planning of hunting strategy, and the possibility of discusing the result, and the establishment and maintenance of relationships with neighbouring bands, would probably have necessitated language. Speech would surely have increased the extractive potential of the hunters.

From our knowledge of modern hunter-gatherers we know that band size averages twenty-five persons. Although there is considerable variation in this figure, depending on the game, the environment, and the technological level of the hunters, it gives some indication of what we might expect. The cave was extremely large, and might well have held many more people than this, but it is unlikely that a very large band would have existed or settled in one cave. The evidence from other such sites again suggests an average group size of twenty-five individuals.

From the skeletal evidence we know that 40% of the individuals died below the age of fourteen; before they reached reproductive age. This suggests that the average woman bore at least four children, for this would allow two to survive to maintain a stable population. From our knowledge of fertility today, it seems possible that more children would have been born than this number and that the mortality of babies may have been greater than the known skeletal remains suggest. Because we believe that prehistoric populations were fairly stable in density (for rapid population increase probably did not pre-date permanent settlements), we can be sure that where natural factors such as disease or predation did not operate to control the population increase, social behaviour in one form or another did so; and as we have seen, this may well have taken the form of infanticide or abortion.

The inhabitants of Choukoutien were undoubtedly more effective hunters than were their predecessors. The climate in central China was more rigorous than that of the African savanna and required the use of skins and hides as clothing, as well as the controlled use of fire, to maintain body warmth. Herbivorous mammals have learned to survive temperate winters either by increased feeding activity, by making stores of food, or by hibernation. Humans survived by living in caves and maintaining a permanent fire, together with regular hunting activity. (Eventually, they learned to store nuts and other food which was slow to decay.) It was the control of fire and the maintenance of body temperature by the use of clothing, however, that allowed these early humans to colonize the cool northern latitudes of Asia and Europe.

THE IROQUOIS

Because of its suitability for agricultural development, there are no hunters and gatherers left that still occupy the temperate biome of deciduous forest. To describe such people, it is therefore necessary to use historical data supplied by western European explorers and colonists. For purposes of comparison with the people of Choukoutien, I have selected the aboriginal inhabitants of New York, the Iroquois, for though they practised agriculture to increase their vegetable food resources, they were still very dependent on gathering wild foods and

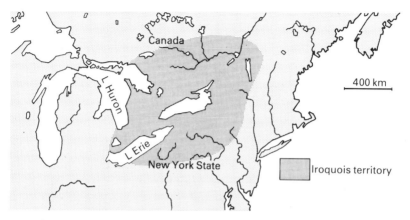

Fig 4.9 The Iroquois occupied a large area of what is today New York State and Quebec, an area rich in deciduous woodland, lakes, and rivers.

hunting. The five tribes that are known collectively as the Iroquois occupied the major part of present-day New York State (Fig 4.9), an area rich in both quantity and variety of resources[7-12]. Several important rivers dissect the rolling countryside which even today is dotted with forests and which was much more heavily wooded in pre-contact times. The northern part of Iroquois territory consisted of coniferous forest, but elsewhere it was mostly deciduous. Principal forms were oak, chestnut, maple, hickory, elm, and birch; woods in this region are interspersed with grassy clearings and marshes, and there are numerous small streams and lakes. The climate is temperate with a 120-day frost-free growing season, making it suitable for maize agriculture. Soils are well developed and are very well-watered and fertile, especially in the broad valleys of the major rivers.

A biome of this kind with very high diversity indices offered the Iroquois an impressive number of alternatives. Although at the time of contact, agriculture provided the basics of Iroquois subsistence, local resources were also used: roots, stem and bark foods, about twenty varieties of berries, and more than eight kinds of nuts[13]. Deer, bear, and moose were the most important game animals, although smaller forms such as beaver, raccoon, rabbit, and squirrel were also considered good eating. Birds, and particularly migratory birds, were hunted in spring and autumn, especially ducks, geese, and passenger pigeons.

Among the Iroquois, agriculture consisted of true field-cropping with villages growing up to several hundred acres of cultivated maize (17 varieties), squash (8 varieties), and beans (10 varieties). Tobacco, artichokes, melons, and sunflowers were also grown but in smaller quantities. Most of the agricultural crops were processed for storage, and some villages stored enough maize to last for several years. Maize was parched and beans and squash were dried[14].

Fishing was an important supplement to the Iroquois diet, especially in the early spring when stored foods might be in low supply. Salmon and eels were the principal catches, and bass and carp were also taken. Fish were either harpooned, shot with bow and arrows, or trapped. Despite the natural richness of the environment, early explorers frequently mentioned hard times when food supplies ran low[15].

The limiting resource among the Iroquois appears to have been game, especially deer, bear, and moose). These animals provided not only the meat in the diet but also the raw materials for clothing, bone for implements, sinew, etc. While beans were a limited source of protein, animal food provided much more, as well as hides for maintaining warmth. We shall return to the importance of hunting below, and examine the effect of this limiting factor on Iroquois social organization.

During the growing season (from May to September) the Iroquois lived in villages that ranged in size from hamlets of 4-5 lodges, or longhouses, to large settlements of over 100 lodges. Village sites were usually located on hilltops and were close to both water and arable land. Even before contact, villages were often surrounded by log palisades and earth ramparts. The lodges consisted of compartments for nuclear families on either side of a long corridor; fires were built in the corridor and were shared by the families on either side. Each compartment contained a sleeping platform and food-storage areas. These longhouse villages were inhabited full-time during the planting and harvest season, but in the late autumn the Iroquois split up into family groups for hunting.

Village sites and their surrounding fields were cooperatively cleared by the men who girdled the trees, burned them, or let them rot, and then removed the roots. Natural clearings were used where available. Cleared land was held in common, and each individual could cultivate as much land as he chose. In the early spring women began the planting operations by hoeing the ground with antler, wood, or bone hoes, forming little hills into which the seeds were placed. Fields were hoed several times during the summer to remove weeds. In autumn when the corn was ripe, the ears were picked and the stalks left standing. The fields were left as they were until the following spring, and the same fields were used repeatedly until the soil was exhausted. Little hunting was done during spring and summer, except for some trapping of local game. Fishing was a major spring activity, however, and was conducted by both men and women. Fish were dried and stored for future use. Iroquois women and children carried on extensive gathering during summer and early autumn—wild fruits and nuts were collected and stored.

After harvest the villagers split up into family groups and hunted from October to January. These small groups travelled considerable distances and lived for the few months of late autumn and early winter in temporary shelters near the areas being hunted. Generally hunting of deer was done singly by Iroquois men, but women accompanied them to the hunting grounds to cook, prepare the game, and carry it home. Deer were generally taken when they gathered in the autumn to feed on nuts. After January bear were hunted during their hibernation period. By mid-winter the small groups returned to the villages with meat and hides. The time between the end of the hunt and beginning of spring planting was used to manufacture the products made of hide, bone, antler, and sinew.

In mid-March there was generally another exodus from the villages. If food shortages were felt, they generally occurred at this time of the year when survival depended on having a good supply of stored food. Expeditions were formed to fish for salmon, which came up the rivers to spawn, and to collect birds' eggs. Sugar maples were also tapped in early spring, and the sap made into syrup and sugar. In May the fishing and sugaring parties returned to the villages and prepared for planting.

The Iroquois had a well-developed technology that was geared to the requirements of their economy. Wood was plentiful and was used for a great variety of objects—from tubs and bowls to their longhouses. Many implements for working wood demanded harder material, and bone and antler were used to make knives, needles, punches, and so on. Metal-working was unknown, and stone was used only for piercing implements such as arrowheads and axes. Pottery containers were made from coiled and fired clay, and baskets were woven from corn husks and rushes. Hides provided much of the material for clothing. Manufacture of these items was divided among men and women—men producing the implements for hunting and fishing, while women made the light tools used in agriculture as well as household goods.

A distinction useful in analysing human technology has been made by Philip Wagner[16]—the distinction between *implements* and *facilities*. Implements are defined as those tools that directly enhance or amplify human effort; examples are spears, levers, hammers, etc. A facility, on the other hand, restricts or prevents motion or energy exchanges without human intervention, and any item that retains heat (clothing, houses, containers) can be classified as a facility. Fish weirs, traps, dams, irrigation systems, would also fall under this rubric. If we examine the history of human technology and humankind's radiation from the tropics into the northern latitudes, we see dependence on a technology with an increasing number of facilities.

When we examine groups such as the Iroquois it is clear that their facilitative technology was quite complex: ceramic containers, clothing, houses, fish weirs, storage pits, etc. Simple machines were also used: bow and arrow, canoe and paddle, and the blow gun.

SUMMARY

As this brief description of Iroquois technology and their annual cycle indicates, the habitat occupied by the five tribes offered them an enormous number of alternatives. High diversity indices of both plants and animals served as a kind of insurance policy in that shortages of one staple could generally be compensated for by a comparable alternative resource in the environment. The marked exception seems to have been the major herbivores exploited by the Iroquois—deer and moose. These animals were essential not only as food but for the hides and other raw materials they provided for the manufacture of implements and clothing. As populations of the five tribes increased, so did competition for available game. Animals that are hunted regularly develop shyness of humans and become more difficult to capture. It should be stressed that the major animals hunted were all *local* game; this means that any over-hunting or shortage of life-space for the game would have had immediate effects on the human groups that depended on them. Undoubtedly the habitat of deer, moose, and bear was adversely affected by the Iroquois land-clearing practice as well as by their custom of using land until it became exhausted. In the absence of domesticated animals and given the fact that the game exploited were in a sense competing with the Iroquois for space, the causes for inter-group rivalry, for which the Iroquois became famous, can be discerned.

Some anthropologists have argued that the militarism and competitiveness of the Iroquois and the consequent formation of the League of

Fig 4.10 The fire drill is one of the most widely used pre-industrial means of making fire among tribal societies, and the technique probably has a considerable antiquity. It is found among many surviving hunter-gatherer groups throughout the world. Here it is in use by San Bushmen (see Chapter 7). (*Irven de Vore, Anthro-Photo*)

the Iroquois were a post-contact phenomenon. However, a number of pre-contact sites are stockaded villages; inter-group feuding and raiding appear to have some antiquity. The evidence suggests that competition was probably over the limiting factor we have discussed: game[17-19]. This has probably been the case throughout much of human evolution and was possibly true in the adaptation of *Homo erectus* in such temperate areas.

The Iroquois lived in a rich and diversified habitat and exploited resources on several different trophic levels. Their technology was well developed, and they were agriculturalists in the full sense of the word. Huge fields were systematically cleared, planted, tended, and harvested, and an effective storage technology allowed them to maximize the period over which a harvest might be enjoyed. The literature of anthropology and archaeology often treats agriculture as though its practice conferred magical properties on its practitioners, practically guaranteeing rapid population growth, sedentary village life, the growth of class distinctions, and the birth of civilization. The Iroquois example demonstrates that no matter how sophisticated the agricultural techniques, any human population is still limited by the weakest link in its food chain; its spectrum of resources, and a storable surplus of grain does not alone ensure an easy life.

On the contrary, it appears that it may have been the introduction of agriculture among the Iroquois which allowed their population to increase substantially to the point at which game became a scarce commodity. And as we shall see in Chapter 9, agriculture will also have initiated the accumulation of property, and the inevitable resulting inequality of wealth.

Finally, it is fascinating to recognize that all the archaeological evidence points to the fact that the key to the expansion of human populations into the temperate zones was their ability to harness and control fire (Fig 4.10). This was one of the most outstanding leaps in humankind's conquest of nature: it not only opened up vast areas for exploitation, but would eventually make possible the use of both ceramics and metals, through the processes of firing and smelting. These were developments which would give their inventor dominion over the planet Earth, and the possibility of exploring the farthest limits of the Solar System.

References

1   Chardin, P. Teilhard de and C. C. Young 1929 Preliminary report on the Choukoutien fossiliferous deposit. *Bulletin of the Geological Society of China* 6: 173–202.
2   Black, D., P. Teilhard de Chardin, C. C. Young and W. C. Pei 1933 Fossil man in China: the Choukoutien cave deposits with a synopsis of our present knowledge of the late Cenozoic in China. *Memoirs of the Geological Survey of China*, Series A, 11: 1–166.

3   Hsu, J. 1966 The climatic conditions in North China during the time of Sinanthropus. *Scientia sinica* 15: 410–414.

4   Weidenreich, F. 1943 The skull of *Sinanthropus pekinensis*: a comparative study on a primitive hominid skull. *Palaeontological sinica*, N. S. D, 10: 1–484.

5   Chaney, R. W. 1935 The occurence of endocarps of Celtis Barbouri at Choukoutien. *Bulletin of the Geological Society of China* 14: 99–118.

6   Breuil, H. 1939 The bone and antler industry at Choukoutien Sinanthropus site. *Palaeontologica sinica*, N. S. D, 6: 7–41.

7   Morgan, L. H. 1962 *League of the Iroquois* (New York: Corinth Books).

8   Murdock, G. P. 1934 The Iroquois of Northern New York. In *Our Primitive Contemporaries* (New York: Macmillan & Co.)

9   Ritchie, W. A. 1953 *The Iroquoian Tribes*. Educational Leaflet #7, New York State Museum.

10   Drumm, J. 1962 *Iroquois Culture* Educational Leaflet #5, University of the State of New York at Albany.

11   Whallon, R. 1968 Investigations of prehistoric social organization in New York State. In *New Perspectives in Archeology*, S. R. Binford and L. R. Binford, eds. (Chicago: Aldine Publishing Co.).

12   Speck, F. G. 1945 The Iroquois: a study in cultural evolution. *Cranbrook Institute of Science, Bulletin* 23.

13   Parker, A. C. 1910 Iroquois uses of maize and other food plants. *New York State Museum Bulletin* 144, No. 482.

14   Waugh, F. W. 1916 Iroquois foods and food preparation. Canada Department of Mines. *Geological Survey, Memoir* 86.

15   Sites, S. H. 1905 Economics of the Iroquois. *Bryn Mawr College Monographs*, Volume I, No. 3.

16   Wagner, P. 1960 *The Human Use of the Earth* (New York: The Free Press).

17   Quain, B. H. 1937 The Iroquois. In *Cooperation and Competition among Primitive Peoples*, M. Mead, ed. (New York: McGraw-Hill).

18   Snyderman, G. S. 1948 Behind the tree of peace: a sociological analysis of Iroquois warfare. *Pennsylvania Archaeologist* 18 (3–4).

19   Hunt, G. T. 1960 *The Wars of the Iroquois: a study in intertribal trade relationships* (Madison: University of Wisconsin Press).

5 The Northern Grasslands and Coniferous Forest

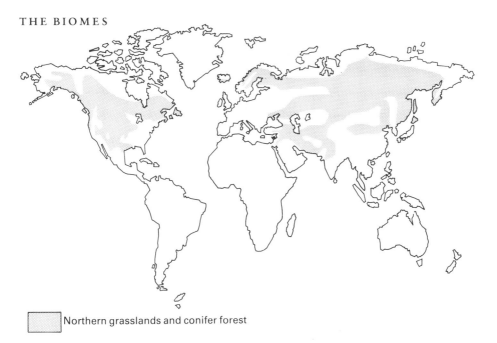

Northern grasslands and conifer forest

Fig 5.1   The northern grasslands are very extensive and very important in the later stages of human evolution when agriculture was being developed, since they constitute the finest cereal growing lands. To the north they are bordered by the far less productive northern coniferous forest which has played only a minor role in the evolution of humankind and human culture.

North temperate grasslands cover wide areas throughout the northern hemisphere; they occur where rainfall is insufficient to support forest and is usually in the region of 250–750 mm depending on temperature and seasonal distribution (Fig 5.1). The grass species themselves vary according to rainfall pattern and temperature, and the flora also includes other herbs such as various daisies and legumes which contain important nutrients for the grazing animals which are characteristic of the biome (Fig 5.2). These are typically the wild horses and saga antelope in Eurasia and the bison and pronghorn antelope in North America. As we shall see, a wider range of species was present during

Fig 5.2   The temperate grasslands and coniferous forest covered vast areas of the northern hemisphere. The grasslands carried immense numbers of grazing herbivores—buffalo in North America and wild cattle in Eurasia—and have since proved to be readily converted into cereal farms, as in the case of the US mid-west prairie farms. The forest regions with their needle-leafed conifers produce dense shade which inhibits herb and shrub growth, but they do generate seeds which are exploited by animals such as squirrels, siskins, and crossbills. (*Natural History Photographic Agency*)

the Pleistocene. Like the much more numerous inhabitants of the savanna, these animals form herds and regularly undertake long migrations which enable them to avoid overgrazing and the seasonal shortages of grass. Burrowing rodents, such as ground squirrels, prairie dogs, gophers, and moles, are also important. The wetter grasslands have recently been extensively ploughed as farm land for cereal growing (an alternative grass species) and the dryer areas have been used for ranching cattle (in the United States) or sheep (in Eurasia). Overgrazing of the arid grasslands has had an effect on the entire biome and in places reduced it to near desert (see Chapter 8).

As we shall see, the temperate grasslands have played a most important role in the later phases of human evolution, but the neighbouring biome, the northern coniferous forest, was of far less significance.

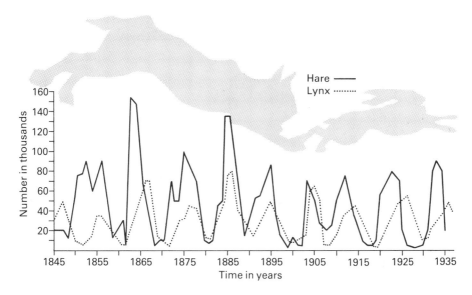

Fig 5.3   Low diversity ecosystems are less stable than high diversity systems, and
their populations of animal species tend to oscillate. This diagram records
variations in the abundance of the snowshoe hare and lynx, its predator, in the
area of the Hudson Bay in Canada. Every 9–10 years the populations drop to a
very low level. This phenomenon would place enormous stress on any predatory
human population which did not have a very diverse diet.

Coniferous forest is characteristic of wide areas of northern Eurasia
and North America and is typified by the presence of extensive stands
of pine, spruce, and fir (Fig 5.2). These large coniferous trees keep the
forest floor in shade, and the lower storey vegetation tends to be poor,
with few shrubs and herbs. Evergreen needles are slow to decay, and
this, combined with the impoverished undergrowth, results in the formation of soils that are considerably poorer in humus than are the soils
of the temperate grasslands or forests. Populations of small animals
such as squirrel can survive on the seeds of the conifers, but the larger
animals that exploit the northern forest are dependent also on neighbouring environments; this holds true for the large herbivores, such as
moose, and also for humans.

Diversity indices in the northern coniferous forest are low, and as
is so often the case in low-diversity situations, there are marked oscillations in populations of both plants and animals. The trees themselves
are often the victim of sudden increases in populations of insect parasites, while other animal life in these areas also undergoes cyclical
fluctuations (Fig 5.3). In general, diversity affords protection against

Fig 5.4 The buffalo, moose (above), and saga antelope are typical large mammals of these biomes in North America. (*Natural History Photographic Agency*)

this kind of fluctuation, so that the stability of temperate forest ecosystems is more evenly regulated than that of the northern forest where there are so often only two or three species of trees. The fauna is composed of species such as deer, moose, wolf, snowshoe hare, squirrel, and grouse (Fig 5.4).

The northern coniferous forest is often interspersed with areas of marshes or small lakes, which allow the development of an insect population that plagues both humans and the large herbivores that inhabit the region. In both Old and New World this biome is often bounded on the south by temperate grassland and on the north by tundra. In this chapter we shall examine three human adaptations to northern grasslands and coniferous forests: one a prehistoric population of hunters and gatherers of the early Ice Age of Spain; a second from the south of France; and the third, an ethnographically known reindeer herding group of tribes in the Soviet Union who depend as well on the other adjacent biome—the tundra. We shall consider this biome in more detail in the next chapter.

TORRALBA AND AMBRONA

The archaeological sites of Torralba and Ambrona are situated midway between the present-day cities of Madrid and Zaragoza on the central plateau of Spain (Fig 5.5). The Ambrona River has cut a valley in the plateau, and both sites lie in this valley, Torralba being three kilometres downstream from Ambrona. Both sites contain remains of Pleistocene occupations that date to the Mindel Glacial period, about 400 000 years ago. The hominids responsible for these archaeological remains were probably *Homo erectus*, who ranged over large parts of Europe, Asia, and Africa during this period. However, no hominid remains have been found at these sites.

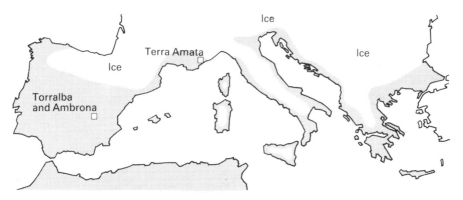

Fig 5.5   The northern shores of the Mediterranean and the southern parts of Spain remained habitable, if cold, during the climaxes of the Pleistocene ice ages, while northern Europe fell under extreme arctic conditions. It was not until less than 100 000 years BP that humans were able to survive in the arctic.

During most of the prehistoric occupations of Torralba and Ambrona the climate was considerably cooler and moister than it is today. The most common tree, as determined by the analysis of fossil pollens, was the Scots pine, with only rare occurrences of birch, alder, and willow. Sedges increased at the expense of grasses during the colder episodes, suggesting a northern coniferous forest interspersed with marshes or alpine meadows. The mean annual temperature was approximately 5°–6° C lower than it is at present.

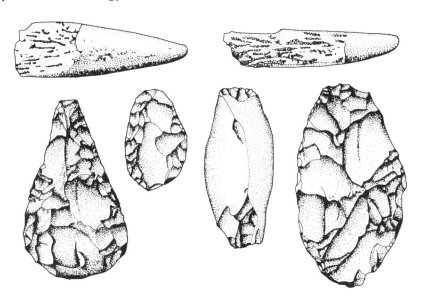

Fig 5.6   Ambrona is important for the richness of its faunal remains and the light that they throw on the hunting prowess of early humans. *Homo erectus* used equipment made from bone, hide, and wood, and their most characteristic stone tools were handaxes. At the top, two pointed tools made from elephant tusks; below, handaxes and scrapers from other European sites of this period.

Neither Torralba nor Ambrona appears to have served as a long-run habitation site; rather, the patterning of stone and bone suggests temporary hunting and butchering stations[1-2]. The game represented are an extinct form of elephant (*Elephas antiquus*), horse, red deer, wild cattle, and steppe rhinoceros. These forms were for the most part not forest dwellers and must have occupied the valley proper and/or the alpine meadows that lay between the pine woods. The numbers of animals indicate repeated use of these locations by early humans. The intermittent marshes and bogs would have been excellent places to trap elephants who could not move away quickly.

At Torralba the excavations conducted by F. Clark Howell exposed some 300 square metres and yielded the remains of about 30 elephants, 25 horses, 10 wild cattle, about 6 rhinoceroses, and 25 deer. The evidence suggests strongly that at least one elephant was killed in a bog and butchered on the spot. In a limited area there are elephant remains consisting of the animal's left side, still lying in place but with most of the right part of the body removed. These bones are associated with four retouched stone flakes. The skull, pelvis, and some of the vertebrae of the elephant were taken away. Not far from this cluster of bones was another concentration, consisting of the same elephant's right humerus, some vertebrae, ribs, clavicle, and maxilla. This area yielded remains of charcoal and two heavy-duty implements of stone (cleavers) as well as some flakes. Howell interprets the large concentration that makes up the elephant's left side as a primary kill area and the second cluster as representing butchering and possibly meat processing.

At Ambrona there are two occupation levels, and it is the lower one, roughly contemporaneous with the archaeological horizons at Torralba, that will concern us here. In this lower level more than 1200 square metres have been excavated. Once again, the dominant species is elephant, and about 30-35 individuals are represented. There are many juvenile animals among these. In contrast with the situation at Torralba, there are no semi-articulated or complete animals at Ambrona; leg bones appear to be missing. One possible explanation is that they were carried away as meat to be prepared and consumed elsewhere. One concentration consists mainly of disarticulated, broken bones along with skull parts. There are only a few artefacts, and these are associated with the broken bones and were probably used to break them up in order to extract marrow. There are several other concentrations of bones and tools at Ambrona, including one that consists of aligned elephant longbones, together with a skull whose cranial vault has been removed, presumably to extract the brain (Fig 5.7). The total tool assemblage from this locality consists of flakes with naturally sharp edges and retouched flakes whose edges have been deliberately modified. These together comprise about 80% of the stone implements, the remainder being larger heavy-duty tools and the cores that are the by-product of tool manufacture. This site has also yielded wood, casts of wood, worked and shaped elephant tusks and the remains of a fire-hardened spear (Fig 5.6).

Although no clearcut structures have been found at Torralba and Ambrona, there are clusters and parts of circles of stones that suggest some sort of tent or shelter. Circles of similar sizes and shapes often mark the remains of tents or huts of ethnographically documented peoples. It has been estimated that the social unit represented by the archaeological data at these sites consisted of a group of about 25, the mean size for known hunter-gatherer groups.

When we compare the remains from Torralba and Ambrona with those of the hominid sites at Olduvai Gorge, we can see that the *Homo erectus* occupants of the Spanish sites were remarkably effective hunters, the densities and gross biomass of game being considerably higher

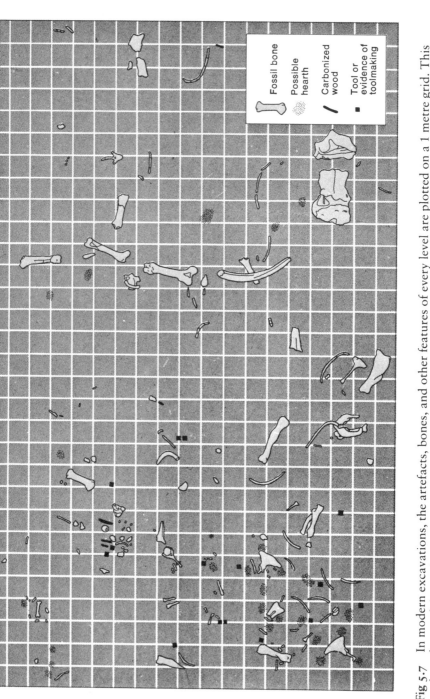

Fig 5·7 In modern excavations, the artefacts, bones, and other features of every level are plotted on a 1 metre grid. This plan of part of the Ambrona site shows the strange alignment of elephant leg bones with a tusk, and to one side a smashed elephant skull. The accumulations of ash and charcoal which suggest hearths lie amongst the bones at the bottom left hand corner. (Based on data from F. Clark Howell, drawing from *Humankind Emerging*, Bernard Campbell (Ed), Little, Brown and Co, Boston.)

than in the earlier Olduvai sites. The number of butchered animals at Torralba and Ambrona suggests that these locations were utilized repeatedly over a period of years for killing and butchering large game. It seems likely that the camps were seasonally occupied, as part of a settlement system involving exploitation of different resources during different portions of the annual seasonal cycle. The sites are located on the fringe of a coniferous forest, and it was the ecotone between the forest and the grassland that would have offered the maximum diversity and quantity of resources. Smaller game would have been available in the forest, while the large herbivores would have been accessible on the grasslands and around the bogs and small lakes that broke up the expanse of pine woods.

TERRA AMATA

A second glimpse of human adaptations during this period has been given us by Henri de Lumley's important excavations at Terra Amata in the suburbs of Nice on the shore of the Mediterranean[3]. Here, during the preparations of the foundations for a new apartment building, stone implements were spotted and building work was halted for a major excavation. The archaeologists sliced through twenty-one metres of the Terra Amata hill to reveal an ancient beach which contained twenty-one levels of prehistoric land surface bearing clear evidence of human occupation, in three different but nearby locations.

One such location was a sandbar, on which four successive huts had been built; on the beach nearby there were six; and on a sanddune a total of eleven huts had been successively constructed—these were of a slightly later date. The latter site was sheltered by a limestone cliff with a freshwater spring nearby: it was an ideal camping site. Here the huts were first revealed by postholes, about twenty-five centimetres in diameter, suggesting that the roof had been supported from inside the hut. The shape of the huts was clear from the bracing stones which lay in a rough oval and the imprints of stakes and saplings implanted close together to form the walls (Fig 5.8). The huts varied in size from eight to fifteen metres long by four to six metres wide. These dimensions suggest that a group of no more than about fifteen people occupied them. In the centre of most of the huts was a hearth—a compact area of baked and discoloured sand, partially surrounded by pebbles to protect the fire from the northwest wind. Near the hearth was a toolmaker's work area with a flat stone as a seat; tools and chips lay

Fig 5.8   The hut at Terra Amata was probably constructed as shown in this drawing. There is not a great deal of difference in the method of construction used here and that employed today (Fig 2.11). The principles may have been understood nearly two million years ago, if the remains at Olduvai have been correctly interpreted (Fig 3.13).

scattered around it. There were also a few bone tools—bones which had been ground to a point—and several pieces of red ochre, worn to a nub, suggesting that the inhabitants had decorated something—perhaps their bodies.

The hut floors also bore traces of animal skins which had probably been slept on, leaving their imprint on the earth. There were also bone fragments—kitchen refuse—and, occasionally, human faeces. The bone fragments suggested that game was abundant in what is now the valley of the Paillon River. The species hunted and eaten included red deer, elephant (*E. antiquus*), boar, ibex, Merk's rhinoceros, wild ox, rabbit, and rodents. There were also remains of seafood: oysters, limpets, mussels, and a few fishbones.

Analysis of the faeces showed the presence of fossilized pollen of a number of spring and early summer plant species, especially broom. It seems that the hunters visited the site regularly in the spring of many successive years, when the wild flowers were shedding their pollen. The climate was evidently cool and moist, and the stratigraphical evidence suggests that the site was roughly contemporary with that of the hunters of Torralba and Ambrona.

In summary, we find at Terra Amata excellent evidence of seasonal campsites, occupied regularly during the spring for a week or so, probably not more. The people who built these huts were evidently migratory hunters and gatherers that made an annual tour of their different seasonal hunting grounds. They tapped a wide range of food resources: vegetables, shellfish, fish, and mammals. Their lifestyle clearly constituted a most efficient strategy, which maximized the available resources of the sheltered valley and coastal grasslands that bordered the coniferous mountain forests of the north Mediterranean coast.

THE TUNGUS

The people referred to here as the Tungus are one of several Tungusic-speaking groups in northeast Asia. Their subsistence is based on hunting and reindeer-herding, an example of transhumance (Chapter 8), exploiting the northern forest during the winter and the tundra during the summer[4]. They occupy the area of the USSR known as Transbaikalia, especially the northern portion, below the 55th parallel (Fig 5.9). The Tungus and their reindeer herds maintain a mutually advantageous relationship: the reindeer provide the Tungus with milk and its by-products and with a reliable form of transportation, while

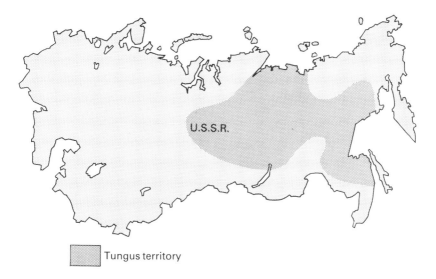

Tungus territory

Fig 5.9   The Tungus people occupy a large part of Siberia, moving as they do between winter and summer grazing for their reindeer flocks. The adaptation of transhumance may date from over 15 000 years BP, but such early archaeological evidence of it is not yet known.

the Tungus afford the reindeer protection against wolves and the voracious swarms of insects that attack the reindeer during the brief summer[5-6].

The winters are long and severe and for their duration the Tungus live in small groups consisting of a few nuclear families. Each family group has its own skin tent which is easily portable, and their mobility tends to be high in order for them to obtain enough food for the reindeer herds. The animals paw through the snow to consume lichen, and since the animals pack the snow down with their hooves, making the moss inaccessible after a short stay, new forage areas must be constantly sought. This scarcity of adequate food for the herds and its resulting mobility keep population density low, on the average about one person per 250 square kilometres (0·01 per square mile).

The reindeer herds are not normally a source of meat for the Tungus, except as insurance against starvation. Hunting and trapping provide meat for the diet as well as skins to trade for guns and tobacco. Deer, bear, elk, and wolves are hunted, while the smaller animals with economically important pelts (squirrel, fox, and sable) are trapped. In this harsh environment the extended hunting range made possible by reindeer transport gives the Tungus an advantage without which their survival would be severely jeopardized. The extreme importance of reindeer to the Tungus is clearly reflected in their affectionate and solicitous treatment of the animals.

When summer comes, the Tungus move out on to the tundra where the reindeer can feed on the few grasses that grow there as well as on a more ample diet of lichen, willow shoots, and reeds (Fig 5.10). With the arrival of the warmer weather the top few centimetres of the permanently frozen ground thaw, creating huge areas of marsh. The marshes afford excellent breeding grounds for flies and mosquitoes, and the Tungus keep smoky fires going most of the day so that men and reindeer are afforded some protection against the swarms of insects.

Despite insects and other hardships, the summer life on the tundra is easier than the winters in the forest and allows for larger aggregates of humans and animals. The summer settlements are not only bigger but tend to be more permanent than the highly mobile winter camps; the lodges are made of birch bark and fires are kept burning continuously to repel insects. Summer settlements generally consist of those households making up a clan. Normally two clans practice reciprocal exchange of daughters in marriage, and it is during the summer that such arrangements are usually made.

Fig 5.10   The Tungus are physically and culturally adapted to the dry cold of Siberia. Only people with advanced technology and with a close relationship to a large mammal species could possibly survive in such a low diversity terrestrial environment. (*Novosti Press Agency*)

Tungus rules governing division of labour and property tend to be highly flexible. Men are often absent for extended periods on hunting trips, but on their return they often lend their help to what may formally be considered female work. The scarcity of food resources is such that strict ownership of game or reindeer would undoubtedly work extreme hardship in many cases. Families generally follow a regular route in their seasonal round of activities, and transportation is eased by their practice of leaving food, clothing, and hunting gear in storehouses for future use. Any relative, however, has access to these goods if he needs them. This kind of flexible reciprocity is highly adaptive in situations of cyclical abundance and scarcity. The size of herds is also subject to fluctuations, and each summer there is a redistribution of animals so that they can be shared equably. Both herds and hunting territories belong to the clan as a whole, even though individual families may exercise stewardship over particular resources.

It is highly improbable that the Tungus could adapt successfully to the harshness of life in the northern forest and the tundra without the added margin of subsistence security afforded by their reindeer herds. It should be emphasized that these herds are fully domesticated animals, not just tamed ones. This means that they constitute a breeding population controlled by humans, and humans can select those qualities of gentleness and docility that make possible the symbiotic relationship between human and reindeer. Skins for clothing, milk for fat and protein, in a country where hunting is unreliable, and transportation over difficult terrain, are all made possible by the harnessing of animal energy by the distinctive adaptation known as domestication (see Chapter 8).

SUMMARY

Because of its low diversity index combined with the wide fluctuations in natural resources that this implies, the northern coniferous forest is a risky habitat for humans. The technologically simple life we see in the remains at Torralba and Ambrona means that people were at this juncture still largely dependent for food on the direct exploitation of large herbivores outside the forest. Only with a new kind of relationship between human and animal that allows for greater mobility can hunter-gatherers survive in the boreal forest and the adjacent tundra.

The distinction drawn in Chapter 4 between implements and facilities is relevant to an analysis of both the Tungus and the inhabitants of Torralba and Ambrona. Certainly without skins to serve as clothing humans could not have survived the rigours of glacial climates in Europe during the Pleistocene, and their successful adaptation to subarctic environments is based on the development of items of clothing, containers, and structures that effectively limit the dissipation of heat, which is a form of energy. The direct evidence for facilities at Torralba and Ambrona is limited to patterning of stones that suggest some sort of tent or other structure. But we do have direct evidence of structures from this approximate time level at Terra Amata and other sites in France.

Certainly the rich technology of the Tungus, together with their use of domesticated animals, renders their adaptation to the northern forest quite secure. In both periods examined in this chapter, the northern forest biome could have supported only very sparse populations, although the Tungus, with their more sophisticated technology and their reindeer, are surely capable of supporting denser groups with less population fluctuation than were the people of the Middle Pleistocene. It seems possible, however, that there was more game on the grasslands during the Pleistocene than there is today, and the adaptations of the people of Torralba and Ambrona may well have brought them rich rewards in mammal meat throughout the year.

References

1   Howell, F. C. 1965 *Early Man*. Life Nature Series (New York: TIME Inc.).
2   Howell, F. C. 1967 Recent advances in human evolutionary studies. *Quarterly Review of Biology* 42: 471–513.
3   de Lumley, Henri 1969 A palaeolithic camp at Nice. *Sci. Amer.* 220, No. 5.
4   Service, E. 1963 *Profiles in Ethnology* (New York: Harper & Row).
5   Hyatt, G. 1919 Notes on reindeer nomadism. *Memoir of the American Anthropological Association*, Vol. 6, No. 2.
6   Mirov, N. T. 1945 Notes on the domestication of the reindeer. *American Anthropologist* 47.

6 The Tundra

Tundra

Fig 6.1 The Tundra circles the north Pole and covers an enormous area of land. The temperatures are very low and the growing season no more than 60 days. The ground is permanently frozen (permafrost) except for the upper few inches during the summer. The vegetation consists of lichens, grasses, sedges, and dwarf shrubs. Tundra arctic food chains were discussed in Chapter 1: this low-diversity biome is very unstable.

Those arctic regions that lie between the northern limit of trees and the southern boundary of perpetual ice are called tundra (Fig 6.1). The tundra biome is wet arctic grassland, and is characterized by extremely low temperatures and very brief summers. These factors, in combination, produce the phenomenon known as *permafrost*—deep zones of permanently frozen ground only the top few inches of which thaw

Fig 6.2 This barren landscape is typical of the tundra. The strange patterning on the surface is due to the presence underground of large polygonal wedges of ice which form the permafrost. Plant life consists mainly of lichens. (*A. Christiansen, Frank Lane Agency*)

during the short summer. The growing season for plants is normally limited to about sixty days. Tundra plants tend to be of small size, and the characteristic forms include lichen (*Cladonia*)—actually the symbiosis of algae and fungi—dwarf willow, sedges, and grasses. Because of the prevailing low temperatures, decomposition is a very slow process and there is little soil formation (Fig 6.2).

Despite the severe limiting factor of low temperatures for most of the year, there is no shortage of life on the tundra. However, because of the extremely short growing season, the pattern of life differs markedly from that in more temperate regions. The large tundra herbivores are present only seasonally, and there are low diversity indices of both plants and animals. The principal herbivore of the New World tundra is the caribou (*Rangifer caribou*) while on the Old World tundra the closely related reindeer (*Rangifer tarandus*) plays a similar role. The seasonal movements of these herds have had a profound effect on the

Fig 6.3 Tundra species can reach very high population densities, but since populations are unstable, dependence on single species as food resources is risky, unless transhumance is practised. Eskimos always include marine food resources in their diets. Left, ptarmigan; right, lemming. (*Natural History Photographic Agency*)

way in which they have been hunted by humans. Migratory birds of many species use arctic areas for breeding grounds, and they too are numerous during the brief months of summer. The typical tundra fauna includes reindeer or caribou, musk ox, arctic hare and fox, lemming, ptarmigan, and snowy owl (Fig 6.3). The warmer months also see huge numbers of blackflies and mosquitoes. It is during these summer months that the Tungus, whom we discussed in the last chapter, move with their reindeer herds from the forest into the tundra while the reindeer graze on the lichens (known as reindeer moss) and other arctic plants.

The low diversity of plant and animal species means that there are more extreme fluctuations in numbers of individuals. Lemmings, of course, are a famous example of an animal whose population numbers undergo enormous periodic increases and declines. The birds and carnivorous mammals that prey on lemmings also experience marked population fluctuations.

The numerous aquatic forms of life in tundra waters are crucial to human adaptations in the biome. They have provided people with important foods. Salmon and seal are resources of critical importance to the peoples we shall discuss below. The two human societies to be examined have both made successful adaptations to the rigors of arctic life. The Magdelenians were the prehistoric occupants of western Europe at the close of the Ice Age, about 19000-10000 years ago. The Nuunamiut and Taremiut are two tribes of New World Eskimos, and the kind of cooperation they display has secured the adaptation of both groups.

THE MAGDELENIANS

During the last sub-stage of the final glaciation of Europe the climate was extremely cold. The Dordogne region of France (Fig 6.4), which today is covered with stands of oak and walnut in the hills and where tobacco is grown in the valleys, was an area of tundra. The fauna at this time consisted of typical arctic forms: reindeer, arctic ptarmigan, snowy owl, as well as the extinct woolly mammoth. The severe cold of this period was somewhat offset by the local topography. The Dordogne is an area of deeply dissected limestone plateaus, and the Dordogne River and its tributaries form a complex of valleys that offer distinctive local microclimates. Overall, however, during the millenia in question, full arctic conditions prevailed. Pollen profiles indicate that there were dwarf willows in the valleys, as well as a few sedges; lichen was the other common plant form.

The Dordogne is a region of ubiquitous caves and rockshelters, and these provided prehistoric populations with sheltered life-space and

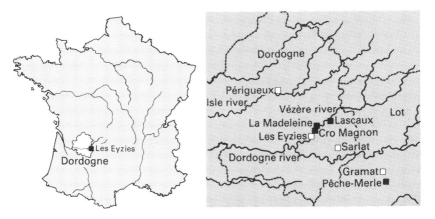

Fig 6.4   The Dordogne is characterized by limestone bedrock deeply incised by the rivers of the region, which hollowed out ledges, shelters, and caves. The valleys provided this shelter, together with game and fish in abundance, and have been inhabited for at least 100000 years.

easy access to water in an environment that was very rich in game. The enormous number of archaeological sites in the area bears witness to the wealth of resources available to these hunter-gatherers. The late Upper Paleolithic populations of the Dordogne, the Magdalenians, developed an adaptation that was highly successful and was in some ways quite similar to that of many living Eskimo groups.

Magdalenian populations were distributed throughout France, as well as adjacent areas of Spain, Switzerland, Belgium, and Germany. The archaeological remains of these people were first documented in the Dordogne, and it is from this region that some of the most complete information comes. The fortuitous combination of naturally sheltered

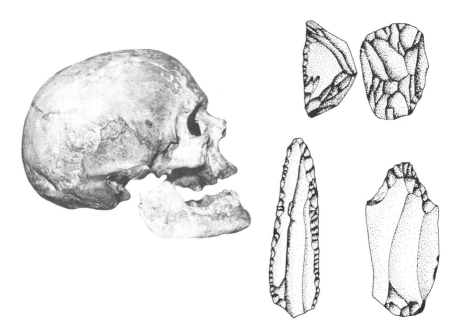

Fig 6.5   The first examples of an ancient race of modern humans were found in 1868 in the deposits of an overhanging rockshelter at Cro-Magnon just outside Les Eyzies on the Vezère river (see Fig 6.4). There were four human skeletons together with flint tools, weapons, pierced sea shells, and animal teeth from necklaces. The stone tools associated with these people were of much better quality than those found in more ancient sites. The skull is that of an old man. (*Musée de l'homme, Paris*)

life-space, facing on narrow valleys through which herds of reindeer migrated regularly, made the Dordogne a favoured area for human habitation.

The humans responsible for Magdalenian archaeological remains were of the fully modern type, most generally known as Cro-Magnon (Fig 6.5). Their skeletal structure is virtually indistinguishable from our own, and they were widely distributed throughout Eurasia by 30 000 BC. People with adaptations analogous to the Magdalenians are well documented throughout the Eurasian landmass during the closing millenia of the Pleistocene, the time that also witnessed the peopling of the New World by the inhabitants of northeast Asia.

The Magdalenian culture is named after the site of La Madeleine in the Dordogne—an enormous rockshelter where the archaeological remains of this period were first recognized. There are now over seventy-five Magdelenian sites known in the Dordogne alone, as well as several in the north of France and in the rolling country between the Dordogne and the Atlantic coast of France. The Dordogne sites occur principally in rockshelters and caves, while those to the west and north are mostly open-air sites, many with traces of cabins or tents. On the basis of this large sample of sites from diverse areas that have yielded a wealth of artefactual material we can reconstruct in some detail the cultural systems existing at this time[1-2].

The Magdalenian adaptation, like that of other cultures from this period, was based primarily on the systematic exploitation of herds of migratory mammals. In the case of the Magdalenians of western Europe, the herds exploited were reindeer. This is not to say that reindeer were the only animals hunted, since remains of bison, mammoth, horse, and many other animals appear in the sites. But frequencies of reindeer exceed those of other animals and often comprise 85–90% of the faunal assemblage; they were the staple of the Magdalenian economy.

In order to understand the nature of the kind of adaptation achieved by these people, a distinction between kinds of resources is useful. *Earned resources* of a given biome are game that gain their energy (food) in that habitat and who spend most of their life cycle within its bounds. *Unearned resources* on the other hand, are game that pass through or spend a portion of their annual cycle in one biome and yet gain most of their energy in other biomes (Fig 6.6). For example, migratory fowl that occur along riverine flyways during the course of their annual travels are unearned resources for the human inhabitants

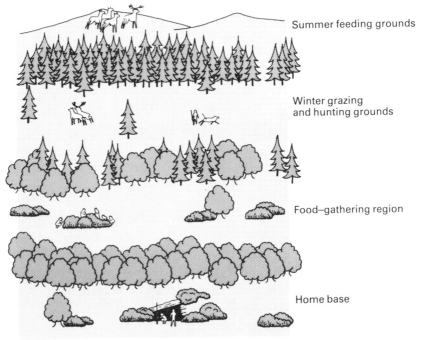

Summer feeding grounds

Winter grazing
and hunting grounds

Food–gathering region

Home base

Fig 6.6 So-called 'unearned' resources refer to migratory herds that live for part of
the year outside the home range of the hunting bands. Herds that feed in the
mountains during the summer may come down to the valleys during the winter
when they are hunted. Thus the valley hunters are drawing on food resources of
the entire region, without actually travelling through it. The figure shows the
relationship of the zones of food production to the home base.

of those river systems, since the birds in question breed, grow, and
mature in other areas. Yet they provide a rich source of food when they
pass through these riverine areas on their way to or from summer and
winter nesting grounds. Fish, such as salmon, that breed in fresh water
but spend most of their life cycle in the sea, are another example of an
unearned resource for riverine populations. Human populations that
live along migration routes of herd mammals—mammals such as wild
cattle, reindeer, or mammoth—and intercept them between their summer and winter feeding grounds are also harvesting unearned resources.

Although the terms *earned* and *unearned* are not entirely appropriate, the distinction is useful, since it allows us to see how energy and
matter gained outside any particular biome can be put to use within it,
thus removing some of the limiting factors within the biome itself. The
implications of this for human technology are important also, since
human groups that are dependent on a resource that passes through an

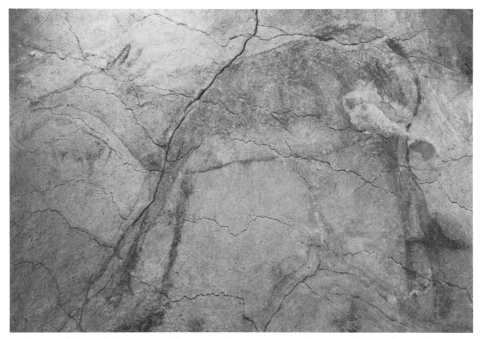

Fig 6.7 The reindeer was of immense importance to these arctic peoples and features in much of their art. It is not known if they were in any sense domesticated. (*Ronald Sheridan*)

environment only twice a year must have some means of extending the usefulness of that resource. In other words, some method of food preservation is necessary for maximum use to be made of the exploitation of such seasonally available resources.

There are some indications that the systematic exploitation of unearned terrestrial resources began just before the initial appearance of anatomically modern people. Certainly the known sites of anatomically modern people from the final phases of the Upper Pleistocene are based on the harvesting of unearned terrestrial resources, most often with marked dependence on one species such as reindeer or mammoth.

The earlier Magdalenian occupations—those dating from about 19 000 BP* through 13 000 BP—reveal an almost exclusive dependence on terrestrial resources, mainly reindeer (Fig 6.7). When climatic conditions became more severe, as they did toward the end of this early Magdalenian period, the inhabitants of the Dordogne began the systematic hunting of new kinds of unearned resources—migratory birds and aquatic mammals and fish. The significance of these additions to their food supply can scarcely be overstated. The following paragraphs

* BP = years before present.

will help to explain why these added foods were so important and will also offer the reader an example of how change in the relationship of just two systemic components—humans and aquatic resources—can have far-reaching effects on the entire ecosystem.

Among ethnographically documented hunter-gatherers early spring is often a period of great hardship. Wild fruits, nuts, and meat such as deer are plentiful in late summer and early autumn, and these are often processed and stored for winter consumption. If supplies run low by spring, the plants that human groups utilize have not yet borne fruit. If the populations in question depend exclusively on one species of migratory mammal, they must wait for it to put in its spring appearance. Any deviation either in the time at which the herds appear or any shortage in their numbers produces disastrous results on the human groups in question. Thus the early spring with its attendant risks would appear to be the period that limits population growth in hunting groups that depend solely on terrestrial resources. If this is true for human groups occupying temperate biomes, the limitations are even more stringent for tundra-dwellers, since these regions are more likely than temperate ones to experience severe fluctuations in animal populations. As we have seen, among many living hunter-gatherers, infanticide, abortion, and other means of population control are rigorously practised. A mother with a small child to care for who must break camp frequently and carry both her child and household gear is not likely to welcome another infant to carry and care for, and she will have few compunctions about taking whatever means are necessary to limit family size.

By exploiting migratory fish and fowl that appear in early spring, the effects of spring shortages in terrestrial resources can be largely overcome. Such an increase in the supply of locally available food means that populations can be more sedentary, and this may remove much of the necessity for limiting numbers of children. Thus, the systematic utilization of aquatic resources serves to offset the effects of springtime limitations on people that are almost totally dependent upon terrestrial resources.

When we compare the sites of the earlier Magdalenian with those of the later period (13000-10000 BP), we find that the later sites are more numerous, larger, and more often situated on river banks, frequently in spots where the rivers narrow. Many of these sites have yielded evidence that they were inhabited all year round. Thus we find that later Magdalenian populations increased in number, and local group size also increased: changes associated with increased utilization of aquatic resources and more permanent settlement.

Before examining what these sites contained, we shall explore some of the implications of the change in adaptation and consequent population growth that occurred between the earlier and later Magdalenian. The model for social organization that is arrived at can then be tested against the data provided in the archaeological record.

The late Magdalenians, like other Upper Paleolithic populations, derived most of their food from exploiting herds of migratory mammals that regularly passed through their environment. In addition, as we have seen, they were dependent upon highly seasonal riverine resources to carry them through the lean months of early spring. The fact that their livelihood derived from seasonally available animals meant that game of all kinds must have been processed for storage and redistributed among groups. The kind of labour force that is needed to hunt successfully and process enough game to sustain a group through several months exceeds that available from a small local band. The 20–30 member group that characterized much of human prehistory was undoubtedly modified with increased dependence on seasonally available unearned resources. At least during the hunting season, much larger groups must have come together, since if we count females, aged and very young males, and also reckon losses due to death or illness, a group of, say, 25 could not have provided more than a few hunters. Aggregation of much larger groups would have been an essential requirement, at least seasonally.

Dependence upon unearned resources, particularly the kinds that were available in the Dordogne, also played a determining role in site location. With heavy dependence on reindeer, sites tend to occur in the narrower river valleys. Undoubtedly this distribution of sites is due, at least in part, to the fact that herds are more easily ambushed in a more confined space. When riverine resources assumed a significant role in the diet of the late Magdalenians, the problem of site location became even more acute. Salmon are most easily captured in the narrows of streams, and such localities would have been most desirable: the type-site of La Madeleine is situated in a narrow valley and at a point where the Vézère River is quite constricted (Fig 6.4). Since such locations are few in number, we can infer that they would not have been casually abandoned when the salmon runs were over each spring. There would have been a distinct advantage in maintaining such a site for use during the reindeer migrations, and especially during the salmon spawning season.

Fig 6.8 Excavation of a Magdalenian site at Plateau Parrain (Dordogne) yielded evidence of a Magdalenian tent. Here the traditional type of shelter is adapted to arctic conditions where very effective insulation is required, and brushwood is not readily available.

Investment in a particular locality by a group and dependence on the resources available in that locality provide the context in which some sort of stewardship might have been a useful social institution. A person, or a particular small group of people, represents the interests of the larger society and in return for status and privileges serves as the representative of the corporate group. Such stewardship involves social distinctions between group members, and these distinctions are reflected by both differential access to goods as well as by symbols of social status. The regularization of such a role, along with rules of inheritance for it, assures the larger society that resources will be looked after and the end-product of the hunt will be shared among the members[3].

Whether or not the late Magdalenian society was a true 'chiefdom' is problematical, but the kind of economy practised would surely have required social distinctions being formalized. This is a new level of sociocultural complexity compared to the essentially egalitarian local band which we believe characterized most hunter-gatherers. Such a society owes its peaceful continuance to the regular means by which seasonally available resources are distributed and the localities for their

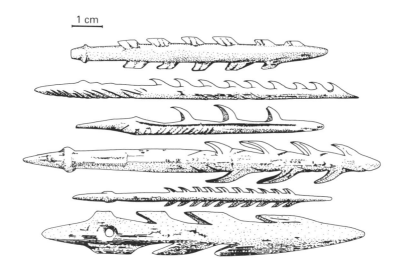

1 cm

Fig 6.9 The Magdalenians developed the art of fishing with sophisticated tackle. This selection of beautifully made barbed harpoons is made of mammal bone.

exploitation maintained. We would, therefore, expect to find in the late Magdalenian the appearance of social symbols and evidence for differential treatment of individuals, with corporate investment in maintaining these distinctions. The elaboration of social symbols is often accompanied by elaboration of ritual, and we might expect increasing evidence for such behaviour in the form of special locations where rituals were conducted.

Evolutionary change in a system involves not only an increase in number and specialized function of components but also new means of integrating these. The kind of social systemic change that we propose for the late Magdalenian represents a major evolutionary step over the kind of band organization that characterized human groups for much of prehistory.

What kind of support can we find in the archaeological record for the deductions made above? The artefactual remains of the late Magdalenian clearly reflect an increased dependence on riverine resources. In addition to the complement of stone tools, which include a variety of specialized implements for piercing and scraping hides and high frequencies of tiny, finely-worked bladelets (that presumably were hafted in bone or ivory to make composite tools) there is a sharp increase in harpoons, leisters, and fish gorges (Fig 6.9). These bone,

Fig 6.10   Spear throwers improved the leverage of the arm on the spear to give a more powerful thrust and a greater range, so that hunting became more often successful. Here a spear thrower is being used by an Australian aborigine to catch fish with a leister or 3-pronged fish spear (*Axel Poignant*).

antler, and ivory implements are finely made and often decorated with both abstract designs as well as representations of salmon and seal. Reindeer hunting was made more efficient by the use of the atl-atl, or spear thrower (Fig 6.10). This is the first simple machine to appear in the archaeological record, and its use added both speed and accuracy to the throwing of a spear.

The evidence for a more sedentary life and consequent population growth is seen in the number, location, and faunal remains of late Magdalenian sites. The sites of the late Magdalenian are three to four times more numerous than those of the earlier Magdalenian, and many locations were utilized that had never been occupied before. It has been estimated that during the period from early to late Magdalenian the population of France more than tripled.

Population growth on this scale apparently had profound effects on the kind of sociocultural systems required to integrate local groups. It is during this period of rapid population growth that there was a

Fig 6.11   Paintings of horses in the Pêche-Merle cave in the Dordogne combine the animals with two kinds of symbol which may be related to hunting magic. The dots may represent projectiles, the handprints surrounding the horses, a person's power over his prey. Evidence for the ritual nature of cave art comes from many sources and especially the fact that much of it is placed in relatively inaccessible places in caves deep in the earth so that it hardly conforms to our ideas of merely decorative art. (*Jean Vertut*)

flowering of art and the use of symbols of social status. Perhaps the most famous of archaeological remains from this period is the exquisite cave art from the Franco-Cantabrian region. Altimira, Lascaux, and many other locations bear witness to the increased importance of ritual in the lives of these people (Fig 6.11). Unless we are willing to assume that humankind's spiritual life suddenly flowered for no apparent reason, we must assume that these famous painted caves (which were not habitation sites) were functionally linked to other components of the Magdalenian cultural system. The use of art and ritual to bind together groups that share a common economic base and to validate the status of individuals to whom the corporate group assigns a leadership role is common among ethnographically known groups.

Corporate recognition of social differentiation can also be seen in the mortuary ritual of the Magdalenians. Human burials sometimes contain elaborate grave goods, including items that must have been obtained through corporate effort, such as thousands of shells from the Mediterranean. Drilled animal teeth, red ochre, stone tools, as well as

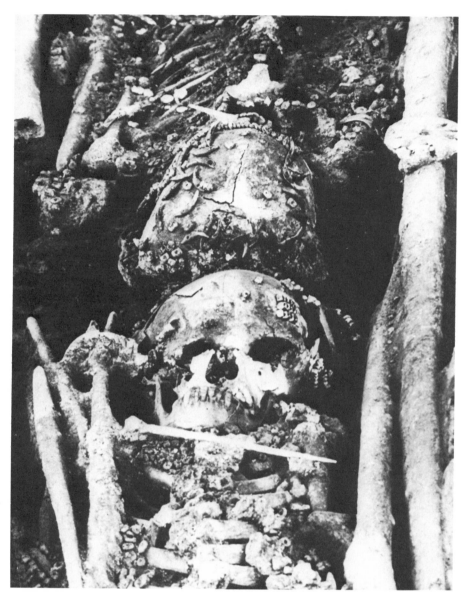

Fig 6.12   This photograph shows part of a burial excavated at Sungir, near Moscow and dating from 23 000 BP. The grave contained two boys, aged nine and twelve years, laid head to head. They were fully dressed and each wore necklaces of mammoth bone beads and the teeth of arctic foxes. On the older boy's chest lay a disk of mammoth tusk carved into the shape of a horse, and both boys were equipped with an assortment of ivory weapons such as lances, spears, and daggers. They were clearly both children of high social status. (*Novosti Press Agency*)

necklaces, pendants, and bracelets were placed with the bodies. Such treatment signifies not only a belief in life after death, but such elaborate mortuary treatment also symbolizes the status of individuals in life (Fig 6.12).

In attempting to reconstruct the Magdalenian way of life, we have singled out a few variables as having had far-reaching consequences. These are: the exploitation of riverine resources (migratory fish, fowl, and mammals) and the techniques by which the basic staple of the diet (reindeer) was hunted. Both of these would have required the maintenance of favourable locations (narrow passes in valleys and narrows of rivers) which, in turn, demand a kind of stewardship and permanent occupation by at least some segments of the society. The development of a sedentary lifestyle required by this adaptation together with the critical effect of riverine resources in removing a limiting factor when people had been almost totally dependent upon terrestrial resources created conditions which allowed rapid population growth. This growth demanded new means of sociocultural integration and the creation of new roles with status for the members of the social units in question.

The increased technological efficiency of the late Magdalenians and the rapid increase in population undoubtedly had negative effects on some of the food resources. Some anthropologists claim that the disappearance of many animal forms (bison in Europe, and mammoth and mastodon in the New and Old Worlds) was the direct consequence of more effective hunting techniques. The reality of this 'Pleistocene overkill' is very much an open question, but the fact remains that between 10 000 BC and 8000 BC many of the large game animals of both Europe and North America either became extinct or retreated to more northerly latitudes as the glaciers retreated, thus forcing new human adaptations to a post-glacial environment. As we shall see, the influence of humans on the environment as pastoralists and farmers was to be much greater than their influence as hunters; they have caused far more numerous extinctions in more recent times through the destruction of entire ecosystems.

THE NUUNAMIUT

The Nuunamiut and Taremiut are two groups of northern Alaska Eskimos who share a common habitat, participate in similar cultural systems, but who occupy different ecological niches[4-5]. Their trade

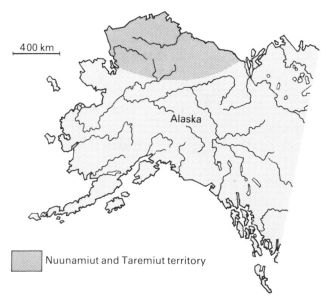

Fig 6.13   The Nuunamiut and Taremiut territory in Northern Alaska comprises
both shore and inland regions, so the people can tap a very broad resource base.
It is now well established that the Eskimo people, with their remarkable cultural
adaptations, entered North America from Siberia about 10000 years ago and
spread throughout the arctic region. Their success depended on their ability to
hunt and kill both terrestrial and marine mammals, as well as birds and fish.

relationships are advantageous to both groups, and they enjoy a kind
of cultural symbiosis that ensures the livelihood of both. The Taremiut
occupy the coast between Point Barrow and Point Hope and their
economic activities consist primarily of the exploitation of sea mammals (whale, seal, and walrus). The Nuunamiut live along the northern
slopes of the Brooks Range and are caribou hunters (Fig 6.13). The
aboriginal population of the Taremiut has been estimated to have been
about 1000, and the Nuunamiut about 3000. Linguistic differences
between the two groups are minimal, and Nuunamiuts can live on the
coast in the summer and participate in Taremiut social life.

   Nuunamiut territory consists of the northern slopes, foothills, and
adjacent plain of the Brooks Range—a rugged chain of mountains some
1000 kilometres long, with altitudes of 2700 metres. These mountains
are crossable only at three main passes: Howard, Survey, and Anaktuvuk. As the northern foothills near the plain their slope is gentle, and
streams become diffuse and sluggish. The arctic plain extends about
100 km from the foothills to the coast and is crossed by many mean-
dering streams; in the summer it becomes a marshland that swarms

with blackflies and mosquitoes. The winters are severe, with temperatures averaging $-34°$ C at the coast and even colder inland. Beginning about November 15th, there are seventy-two days of total winter darkness. The summer is marked by an increase in sunlight, with two months of perpetual light. It is at this time that the top few inches of permafrost thaw and the ephemeral tundra plant life appears. These plants are shallow-rooted and consist mainly of lichens and mosses. The foothills and passes contain dwarf willows and many varieties of berries. The inland fauna are the usual arctic forms, and the summer sees hundreds of thousands of caribou come through Anaktuvuk Pass to feed in the northern plains and foothills.

The main caribou migrations occur in spring and autumn—the herds moving north in March and April and south in September and October. During these migrations large hunting parties assemble, normally about 200–300 people of whom 50–100 are hunters. The autumn settlements consist of family houses and a ceremonial house that is cooperatively built. The Nuunamiut live in these temporary villages from September until early November, then disperse to their winter settlements. During the autumn migration corrals are constructed, and these enclosures are about 45 by 90 metres. Caribou are led into the corrals and either snared or shot with arrows. This is repeated day after day for several weeks, and on a good day 200–300 animals can be taken. The women then skin the caribou, and the meat is cut in strings and dried, or pounded with fat and berries. Leg bones are cracked for marrow, fetuses boiled and eaten, as are the stomach contents. A good deal of this meat is then stored at the campsite.

In November the Nuunamiut disperse into groups of two or three households and settle in the foothills. The standard winter house is the *iceliik*, a round or oval tent-like structure supported by four upright posts with light beams connecting them at the top. Willow branches are attached to form a dome-shaped frame which is then covered with skins. Generally two layers of skins are used, with an air pocket between them for insulation (Fig 6.8). Until March the Nuunamiut remain in their winter settlements, occasionally hunting individual caribou, doing extensive trapping, snaring ptarmigan, and doing some fishing.

In March the scattered groups assemble again for the spring caribou hunt which is similar to that of the autumn. During the spring, ground squirrels, marmots, and grizzly bears are trapped. The end of the spring hunt is marked by feasting and celebration. Many families then leave and go down the Colville River drainage system in skin boats and spend

Fig 6.14   These semi-subterranean sod houses are also built by the Nuuniamut and are very effectively insulated against the cold. The roofs are usually made of whale ribs covered with skins. (*Steve McCutcheon, Frank Lane Agency*)

the summer on the coast fishing, hunting and trapping. It is during this period also that they conduct their trade with the Taremiut. Some families stay behind in the inland area. In August they collect berries and roots, and at the end of summer households again come together for the autumn hunt.

When caribou are scarce the Nuunamiut also hunt mountain sheep, moose, bear, beaver, ground squirrel, marmot, Arctic hare, and porcupine. Wolves, wolverine, and foxes are taken for their skins but are eaten only when other resources fail. While occasional seal are hunted in the Colville River from kayaks, marine mammals are obtained principally from trade with the Taremiut.

Probably the most important supplements to the Nuunamiut diet of caribou are ducks and geese. These appear in the north in early spring and are highly valued for both meat and fat. Ptarmigan, which are snared in late winter, also provide a very important addition to the diet.

Summer fishing provides roughly 10% of the Nuunamiut diet, and the principal means of hunting is by netting or trapping. Plants make up only about 5% of the subsistence base, and these are mainly berries.

Given the rigours of the climate, fat and oil for both consumption and fuel are probably the limiting factor in Nuunamiut population growth, a factor which is somewhat offset by their trade relations with the Taremiut, as we shall see below.

THE TAREMIUT

The principal economic activity of the Taremiut is the whale hunt which takes place in the spring. In late April the baleen whales appear in the offshore ice leads, and whaling camps are set up on the edges of the leads (Fig 6.15). Whaling crews pursue the whales in umiaks with harpoons and attempt to fix as many harpoons as possible in the whale. When the whale rises for air between the ice leads, a lance is used to spear it in the heart or brain. The carcass is then towed toward camp, hoisted onto the ice, and the whale divided by the whole community with great ceremony.

Fig 6.15   Baleen whales (and indeed most other marine mammals) constitute an extraordinarily valuable food resource for the Eskimos, providing as they do, both meat and oil. Their bones are used for the construction of huts and sleds. Here two whales are about to be hauled ashore. (*Steve McCutcheon, Frank Lane Agency*)

A large baleen whale weighs over a ton and is about 18 metres long, providing enormous amounts of meat and blubber. In a good season as many as 35 whales are caught, and the Taremiut make use of every part of the animal. Blubber and meat are eaten raw or boiled, and bones are used to make the frames of houses or sleds. Oil is used for food and fuel. The Taremiut are dependent upon the whale for so many things that in years when the whales changed their migration the Taremiut experienced widespread starvation.

Unlike the Nuunamiut, the Taremiut occupy permanent villages and disperse only in the summer; the villages range in size from 50-300 inhabitants. Houses are circular with extended entry-ways that serve to control drafts and also as storage and cooking areas. In early summer when the whaling activities end, the Taremiut disperse—some families go inland to hunt caribou, others to fishing camps; then the men return to the permanent villages for walrus hunting. Summer is also the period of trading with the Nuunamiut, an activity that involves a long journey to the Colville delta. Trade is crucial to the survival of both groups, and preparation for trading takes place throughout the year.

TRADE AND RECIPROCITY

Trade is conducted through formaliz𝑒d trading partners, an arrangement that has distinct advantages insofar as partners tend to extend themselves on each other's behalf and offer each other generous exchanges. Only after a man completes trading with his partner is he free to trade with others; a wealthy man might have five or six trading partners. As many as 400-500 people assemble at a single location to carry on exchanges.

Table 6.1   List of items traded between the Nuunamiut and Taremiut

Nuunamiut to Taremiut	Taremiut to Nuunamiut
caribou hides*	oil*
wolf, fox, wolverine skins	walrus and seal skins
sheepskins and horns	finished wooden vessels
bones for marrow	stone and slate projectile points
sinew and skin	seal flipper bags for water thawing and storage
pitch for glue	ivory and driftwood
wooden and stone objects	
pemmican and berries	

Note: * These are the most critical items in trading. Data from Spencer[4].

As this list (Table 6.1) indicates, complementary trading offers distinct advantages to both groups. Certainly in the essentially treeless area occupied by the Nuunamiut, and given the rigours of the climate, fuel is a critical resource. Maintaining personal warmth through having the appropriate kinds of hides from which parkas and mukluks can be made is essential to Taremiut survival, and the skins obtained from the Nuunemiut fill this need. However, the principal limiting factor on Taremiut population growth appears to be the fact that the whaling season occurs only in the spring, and then it is brief.

Among both the Nuunamiut and Taremiut, sharing of food within a community is an essential part of survival behaviour. No one goes hungry before the whole group does, and generosity is the basis for prestige. Territory is thought to be common to all, and no single individual or sub-group has exclusive rights over an area's resources. Group affiliation is generally dictated by kin and marriage ties. Usually a wife goes to live with her husband's family, since men are reluctant to break ties within their hunting or whaling groups, but these arrangements are flexible. Among both the Nuunamiut and Taremiut, individual leadership is conferred on men in the context of hunting. The competent and wealthy Nuunamiut hunter is the organizer of the hunt and the one who builds the stockades. The captain of a whaling crew among the Taremiut has no more decision-making power than his crew members, but his own share of the catch is larger. This privilege is offset by his obligation to provide a feast for the community at the end of the whaling season.

Following extensive contact with western cultures, there have been profound changes in these relatively stable adaptations of the Nuunamiut and Taremiut. Snowmobiles have replaced dogs and sleds, and hunting is done with guns. It is now possible for five or six Nuunamiut to exterminate very large portions of caribou herds. Hunting has become so easy that care is no longer taken to utilize all parts of the animals; only the choice portions are kept. They also kill more than they can consume to sell the surplus on the coast.

SUMMARY

One of the striking contrasts between the pre-contact Eskimos and the Magdalenians is the fact that the Magdalenians had neither boats, dogs, nor sleds. In the kind of broken terrain they occupied sleds would have been of little value, but Magdalenian survival without riverine transport is a testimonial to their hardiness and ingenuity. Certainly the arctic regions of the world were not permanently occupied until after dogs became domesticated and showshoes or sleds and boats facilitated transportation over the wide areas of tundra they covered during an annual cycle. Both our stone age and ethnographic examples were dependent upon a fairly high order of technological sophistication—technologies that used many facilities such as ice houses, tailored clothing, traps, and so forth, as well as simple machines such as harpoons and spear-throwers. Unearned resources provided the overwhelming portion of the subsistence base for both Magdalenians and the Eskimos, since the tundra cannot support resident populations of herbivores year-round. Migratory birds and aquatic resources served as important supplements to the diet, and in the case of the Magdalenians the addition of salmon to the diet allowed for rapid population growth and increasing permanent settlement. The rate of cultural development was clearly accelerating and it is tempting to wonder what the Magdalenian culture might have developed into had not the Ice Age come to an end, and the climate and fauna changed drastically. Certainly there was great potential among the late Magdalenians for further cultural elaboration. However, with the retreat of the ice sheets from western Europe, local populations were compelled to make new adaptations to a new, temperate climate.

References

1   de Sonneville-Bordes, D. 1960 *Le Paléolithique Superieur en Périgord* (Bordeaux: Imprimeries Delmas).
2   de Sonneville-Bordes, D. 1967 *La Préhistoire Moderne* (Perigueux: Fanlac).
3   Service, E. R. 1962 *Primitive Social Organization: an evolutionary perspective* (New York: Random House).
4   Spencer, R. F. 1959 The North Alaskan Eskimo: a study in Ecology and Society. *Bureau of American Ethnology Bulletin 171.*
5   Gubser, N. J. 1965 *The Nuunamiut Eskimos: Hunters of the Caribou* (New Haven: Yale University Press).

7 Hunters and gatherers

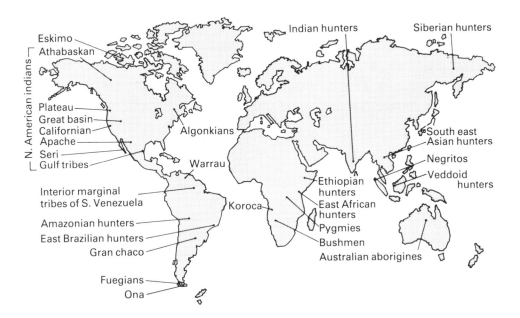

Fig 7.1 Remaining groups of hunters and gatherers survive in areas which have proved unsuitable for agriculture, and therefore marginal as far as human survival is concerned. The fact that today these people can make a good living in these areas is due to their low population density, and their wide-ranging skills. It would follow that in the more productive areas, now given over to agriculture, their lives would have been even less rigorous. Today they are threatened not by the natural environment, but by neighbouring agriculturists and pastoralists, by misguided governments, and by a host of destructive western cultural influences.

THE LIFESTYLE

In the first part of this book we have described a number of living and historical groups of so-called hunter-gatherers. Today the last of these peoples are seriously threatened and survive only in very small numbers. The map (Fig 7.1) shows the location of most remaining groups. As we have seen, none has survived in temperate regions, and those that remain are mostly found in marginal areas in which agriculture has not been developed. The main areas of this kind which are occupied by hunter-gatherers are the tropical rain forests, the near deserts of the tropics, and the arctic. The fact that human groups can adapt to these biomes in areas which are marginal with no metal technology (though today most of them have some imported iron goods) emphasizes the effectiveness of these pre-agricultural and pre-industrial human adaptations.

THE SAN BUSHMEN AND THE SAVANNA

In this chapter, we shall look more closely at the hunter-gatherer adaptation to try to understand in greater detail how they have been able to exploit their environment with such success. At the same time we can get further insight into the ecological adaptations of *Homo sapiens*, when the species was not yet in a position to make major changes to its environment.

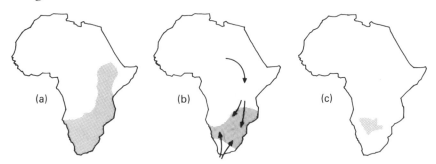

Fig 7.2   Evidence from skeletal remains suggests that the San peoples once occupied a much larger range in both South and Eastern Africa. Linguistic evidence supports evidence of San people as far north as Tanzania (a). In 1652, when Dutch settlers arrived at the Cape, the territory of the San was already reduced in the face of black Congoid people spreading from the North and West (b). Today, following the expansion of Europeans, the San people are reduced to a mere remnant almost entirely confined to the Kalahari desert in Botswana, South Africa, and Angola—an arid area that no one else wanted (c). Though their genes will survive, the San culture appears to be doomed.

Fig 7.3   The San people are friendly, attractive, and generous. A number of recent expeditions have between them made a very detailed study of San society and ecology. (*Shostak, Anthro-Photo*)

The most detailed and revealing examination of a hunter-gatherer group was carried out by a Harvard expedition to Botswana, where small groups of the so-called bushmen, Khoisan or San people of Southern Africa were studied[1-2]. Today only some 55 000 of these people survive in Botswana, Namibia, Angola, Zambia, and the Republic of South Africa (Fig 7.2(c)), but their numbers were far greater in the past[3]. The San are now believed to be the ancient inhabitants of much of Southern and Eastern Africa, and though distantly related to them, they are biologically distinct from their neighbours, the black Bantu-speaking or Congoid people who today occupy most of Africa south of the Sahara. In contrast to the blacks, the San people have yellow skins, with little subcutaneous fat—so their skin wrinkles—small stature, small teeth, epicanthic eye folds, and other distinctive features such as very tightly curled hair (Fig 7.3). Most striking perhaps is the presence of steatopygia often seen in women—the development

Fig 7.4   The San people are small, and their skin a yellow-brown. The women are distinctive in the tendency to develop large fat reserves on their thighs and buttocks. Their faces are of characteristic shape and very handsome. (*Lee, Anthro-Photo*)

of extensive fat deposits on buttocks and thighs. It has been suggested that this is an adaptation to give women protection from occasional famines—a protection they might need during pregnancy.

A closely related group, the Hottentots, are today almost completely absorbed by European and other settlers with whom they have crossed to form the 'Cape Coloured' people of Cape Province. The Hottentots were, however, traditional pastoralists rather than hunters, at least in historical times.

In 1652, when the first Dutch settlers under Jan van Riebeeck reached the Cape Peninsula, the San occupied most of the southern part of Africa (Fig 7.2b). In prehistoric times (c. 10000 BP) it is believed on the evidence of skeletal remains that they occupied a very large part of Africa (Fig 7.2a) though their exact distribution is uncertain. Their language is quite distinct from other African languages as it contains a number of kinds of click sound. Some of these clicks are found today in the language of peoples as far north as Tanzania, and this evidence supports that of the skeletal remains. Taken together, the historical evidence suggests that during the last 10000 years, the San have been driven down into southern Africa by the expansion of the black Bantu-speaking peoples. Eventually they were decimated and expelled from all but the most marginal areas of the Kalahari Desert by the Europeans expanding north and east from the Cape.

Today, as a result, the San are confined to the least productive part of their original range—one of the few areas not suitable for agriculture nor very productive for pastoralism. The Kalahari Desert where they survive is a semi-desert area of sand-veld with an average annual rainfall of 100–400 mm, and years of drought are common. The sand is poorly covered by grass or scrub with scattered thorn trees. Travelling northwards in the Kalahari area the rainfall rises to 800 mm per annum and the desert blends into the woodland savanna. A glance at the map (Fig 2.4) shows that in general the savanna zones of East and South Africa were probably the original homeland of these people, and a very rich and productive homeland it was.

The savanna, as we have seen, with its extensive grasslands, carries a wide range of grazing and browsing species, and this is made possible by the complementary nature of the different species' food requirements. The chart (Fig 7.5) shows clearly how the game in a typical savanna area is distributed across the biome from grassland to forest edge. Even in a particular area, different species tend to eat different grasses and herbs.

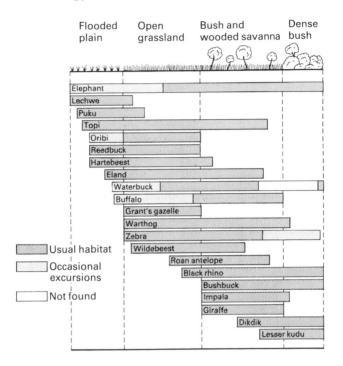

Fig 7.5 Different species of herbivore occupy different parts of the savanna according to the rainfall and its distribution during the year. Grassland is a product of a climate with a long dry season which cannot sustain tree growth. (After Jewell[4])

Research in East Africa indicates that the most important single factor in determining the productivity of the savanna are the mean annual rainfall and its distribution. In wet areas (1000 mm or 40 ins rain) the standing biomass of large herbivores is ten times as great as in dry areas with only 220 mm (8½ ins) rain (see Table 7.1)[4]. From the point of view of the hunter, as for the carnivore, a stable population can survive so long as the ratio of herbivore to carnivore is at the appropriate level. An approximate ratio of one carnivore to one hundred herbivores is typical, but this will vary according to the species of predator and the species of game.

The result of variations in rainfall and productivity for the human hunter is twofold: first the density of the human population of the area will vary, and second the average distance needed to find suitable game will change. It is for this reason that the remaining San, now living in an area of very low rainfall (125–800 mm) do still make a reasonable living by hunting over a large area so long as some game is present. In places where

Table 7.1   Calculated standing
crop biomass of large mammals, production, and rainfall data.

Locality	Large herbivore biomass kg/km²	Production kg/km²/yr (liveweight)	Annual precipitation (mm)
A Wildlife areas			
Bunyoro North, Uganda	13261	1145	1150
Rwenzori National Park, Uganda	19928	2554	1010
Ngorongoro Crater, Tanzania	7561	1503	893
Nairobi National Park, Kenya	4824	1008	844
Serengeti National Park, Tanzania	8352	1743	803
Samburu-Isiolo, Kenya	2018	402	375
Amboseli Game Reserve, Kenya	4848	934	350
B Pastoral Areas			
Kaputei district, Kenya	7884	–	710
Samburu district, Kenya	6514	–	500
Garissa district, Kenya	3818	–	398
Turkana district, Kenya	2406	–	330
Mandera district, Kenya	1901	–	228
Wajir district, Kenya	1151	–	218

After Coe *et al*, 1976 and Kenya government reports.

the game migrates out of the region, following a drought, the people may find it very hard to survive unless they themselves move as well.

SAN SOCIAL ORGANIZATION

The San live for most of the year in camps around a permanent water-hole. During the dry season, the winter, the entire population is grouped round these holes, between ten and forty at any one water-hole. They consist of a family of siblings, their spouses, and their siblings and spouses, together with all their children. They build and live in huts made of branches and grass (Fig 7.6), usually grouped round a central area and dancing ground. They live in these dry season camps for between three and five months a year and rebuild them annually. Hunting and gathering usually take place within a radius of about six kilometres of each camp—a day's walk—under normal conditions. The camp is a self-sufficient subsistence unit.

During the rainy summer season, the group moves away from the permanent water-holes to exploit the resources surrounding the tem-

Fig 7.6   San huts and San lifestyle have probably changed little over the course of hundreds of thousands of years. (*Lee, Anthro-Photo*)

porary springs and water-holes which appear scattered through the region. During this period, vegetable foods are more varied and are easier to find. Temporary camps are set up. After a day's gathering and hunting the food is pooled each evening and divided between the members of the group. There is a strong emphasis on food sharing within the group, but trade in food between groups is rare: each group is in practice self-sufficient.

In contrast, there is much movement of people between groups, and a great deal of visiting between camps. Relations between the groups are friendly, and people move from one group to another frequently, either owing to marriage or in the search for a more productive area. In practice the composition of these groups changes from week to week with the comings and goings of people. The close relationships that result between groups are most important since they act as an insurance during times of drought and food scarcity—rare though they may be. Hunting areas are not demarcated nor defended: through maintaining good relations with its neighbours, each group has access to a very large area—perhaps ten times its normal territory—which allows it to survive the occasional years of serious drought.

Permanent water-holes, essential for a few months each year, may also be shared if, as a result of drought, there are fewer than usual. The water-hole is a key resource for the San, and probably the limiting resource. The ability of groups to share water-holes without social disruption in times of extreme drought is clearly essential for their survival.

In some parts of their range, the San share their land with black Congoid pastoralists (Tswana and Hereros) who live in scattered villages. Although their cattle share the productivity of the area with the people and the wild game, they do not appear to have a significant effect on the San adaptation we are describing nor on the density of the San people. This is most probably because the presence of the cattle does not significantly alter the extent of the limiting resource—perennial water. Relations between the different groups are always cordial.

SAN FOOD RESOURCES

The eating of fresh meat, though it has been an important factor in human evolution, is nevertheless not the most important part of the diet of most hunter-gatherer peoples. The group of San studied in detail by Lee (called the !Kung) consumed a diet of which 33% of the caloric content came from meat and 67% from vegetable foods. The average daily consumption was 2140 kilo-calories (kcals) and 93 grams of protein per person. The US Government Recommended Daily Allowance for people of the size, stature, and vigorous activity of the San is approximately 1975 kcals and 60 grams of protein. On this basis it appears that San food consumption exceeds this recommended allowance by 165 kcals and 33 grams of protein.

The mongongo nut or mangetti as it is sometimes called (*Ricinodendron rantanenii*), which is common in the area where the study was made, develops within an edible fruit and is produced by a drought resistant tree. The average daily consumption of these nuts accounts for fifty per cent of the vegetable diet by weight and thousands of kilos of nuts are harvested and eaten each year. Thousands more rot on the ground for want of picking. The nut is a remarkable food source since it contains five times the calories and ten times the protein of an equal quantity of cooked cereals. The average daily consumption per person

Fig 7.7   Women are occupied gathering fruit, nuts, and edible roots together with other small food items such as eggs and insects. These constitute the staple foods of the San. (*M. Shostak, Anthro-Photo*)

of 300 nuts (weighing only 212 grams) yields 1260 kcals and 56 grams of protein. Owing to the hard outer shell, the kernel is protected from rot on the ground and good nuts can be collected throughout the year. A diet like this is much more reliable than one based on seasonal cultivated vegetables and cereals.

Besides the nutritious mongongo nut, the San also collect about 100 species of roots and bulbs. This large variety allows the San many alternatives in their choice of a subsistence strategy. During the summer months when the rains come, there is a choice of the tastiest vegetables and fruit, which are easy to collect. During the dry season, the diet necessarily changes because the people have to search for roots, bulbs, and edible resins. The less attractive and rarely used vegetables provide a nutritional reserve at the end of the dry season—when calorie intake falls—so that there is always a margin of safety. Recent research showed no evidence of qualitative nutritional deficiency among the San: there was no sign of kwashiorkor—a common indication of protein deficiency among African agricultural peoples, among whom roots, tubers, or bananas provide the main carbohydrate intake.

Gathering of vegetable foods, which constitute 60-80% of the total diet by weight, is carried out by the women (Fig 7.7) who spend the equivalent of two or three days a week at this task. The men, who hunt medium and large animals with bows and poisoned arrows (Fig 7.8) also collect small animals and plants; they, too, spend an equivalent of two to three days a week out of camp, and the whole group spend the equivalent of four or five days a week in camp, gossiping, singing, dancing, and generally relaxing.

It appears, therefore, that these people lead a surprisingly leisured life and have a very good and a very diverse diet. There is seldom more than two or three days' supply of food in camp, and no need for anxiety about the morrow. Food resources are abundant, and reliable. While hunting may be a risky business, because hunters often return from a long day empty-handed, the rewards are great in terms of the value of the meat. Gathering on the other hand is a low-risk, high-return venture; plant food constitutes a reliable and easily found source. In short, these San of the Dobe area of the Northern Kalahari region eat as much vegetable food as they need, and as much meat as they care to get. The people live to a ripe old age with few signs of anxiety or insecurity. They continue to be well-fed when pastoralists sharing the same environment are seriously short of food.

It is certain that some San groups in the more southerly region of the Kalahari Desert have a harder time than the people studied by Lee. The effect of lower productivity in general would be to make it necessary for people to travel further from camp to get the resources they needed. As a result they would cover a larger territory during their year-round activities, and their density would be less. This apppears to vary from 0.1 to 0.5 people per square mile (2.5 km²) in areas occupied by the San.

Fig 7.8   In all hunter-gatherer societies hunting is the preserve of men. It is a more difficult and risky occupation than gathering, but generates food that is highly valued, thus conferring status on hunting prowess and masculinity in general. The division of labour is universal and highly adaptive. (*De Vore, Anthro-Photo*)

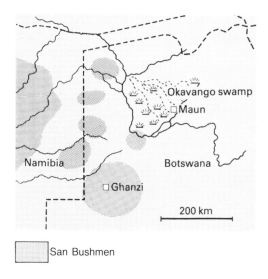

San Bushmen

Fig 7.9 International boundaries in most undeveloped countries were drawn in the distant palaces of Europeans with no knowledge of the geography or demography of the region. The Namibia–Botswana boundary is a line on a map which happens to split the homelands of a number of groups of San: but they are allowed to cross the barbed wire in certain places.

PROSPECTS FOR HUNTER-GATHERERS

In spite of the apparent attractions of the hunting-gathering lifestyle, the San people are today changing to a number of other possibilities. At the turn of the century perhaps 60% were hunter-gatherers; today less than 5% are believed to follow their traditional lifestyle. In Botswana, San families work on the cattle farms, and they have also set up their own communities based on agriculture, and cash game hunting. In the more remote areas, unmarried men work intermittently for the local Tswana and Herero pastoralists in return for subsistence and a goat or donkey. Far more devastating, however, than these developments, is the loss of their traditional lands which are being turned into freehold tenured farms, with the result that they find themselves poachers and squatters on land they have occupied for centuries.

Although there were systematic extermination campaigns of the San during the early part of the last century in South Africa, the most damaging development has been this continual appropriation of their land by both African and European pastoralists and farmers. The land is the essential basis for the hunting-gathering way of life and as they lose it, their traditional lifestyle is increasingly doomed. Even in areas

Table 7.2   Most important foods used by the San

Plants	Animals	
100 edible species, including:	54 edible species, including:	
mongongo	warthog	porcupine
baobab	kudu	leopard tortoise
sour plum	duiker	ant-bear
marula nut	steenbok	francolin
melons	gemsbok	korhaan
roots	wildebeest	hare
	spring-hare	rock python
	guinea fowl	flying ants

After Lee and De Vore[2].

which are otherwise undisturbed, national boundaries were drawn without regard to the peole who live in the area (Fig 7.9).

Recent research into the other remaining groups of hunters and gatherers such as the Hadza (Chapter 3), has shown that the San are fairly typical of these few remaining people. Fifty-eight such societies were examined by Lee[1], and he showed that hunted foods (land and sea mammals) *usually* constituted between 20% and 45% of the diet and gathered food (wild foods, small land animals, and shellfish) the rest. However, in northern latitudes, fishing became extremely important (Chapter 4). Table 7.2 shows the most important foods used by the San. Table 7.3 shows the primary subsistence source of all of Lee's fifty-

Table 7.3   Primary subsistence source, by latitude,
of 58 hunter-gatherer societies

Degrees from the Equator	Primary subsistence source			
	Gathering	Hunting	Fishing	Total
More than 60°	–	6	2	8
50°–59°	–	1	9	10
40°–49°	4	3	5	12
30°–39°	9	–	–	9
20°–29°	7	–	1	8
10°–19°	5	–	1	6
0°–9°	4	1	–	5
World	29	11	18	58

After Lee[1].

eight societies. The few societies in which hunting was predominant were those in arctic and north temperate regions, where, as we have seen (Chapter 6), vegetable foods are scarce. As a result of such comparative studies the flexibility of hunter-gatherer social organization has also become clear. Where food resources are widely scattered, sparse and unpredictable, human groups must be small, well dispersed, and mobile. Where there are large herds of game, and the vegetable foods plentiful, larger groups of people could gather and mobility would be reduced. Changes in the environment were followed rapidly by changes in group size, and these were often of a seasonal nature. Among the (now extinct) hunters of the Great Plains of the American West, the people gathered during the summer into large aggregations to hunt bison, which at that time roamed there in large herds. In the winter, the bison dispersed widely and the hunters did the same[5]. Easy movement between San groups allows rapid adjustment of band size according to changes in productivity of different areas from season to season.

One of the most important questions which has recently arisen in the study of human prehistory relates to the nature and function of territoriality. Investigations among the San indicate no true land ownership and a relaxed attitude to land use by neighbouring groups. Nevertheless among many peoples such as the Australian Aborigines local groups have a longstanding right to hunt and collect in certain areas which they clearly recognize and demarcate. It is difficult to generalize, but it looks as if competition between groups for resources evokes a recognition and demarcation of territory and its identity with the group that lives there. Where it has been recorded, increased competition is usually a result of a reduction in food availability in the area, or follows an increase in population density.

Because many of the remaining hunter-gatherers today subsist in what we would call marginal areas, it seems reasonable to conclude that in the past they usually made an easier living than they do today. This would certainly be true of the Hadza and the San. We know that our ancestors (*Homo*) survived as hunters and gatherers for at least two million years, and the fact that we are here to discuss it suggests that it was an entirely successful adaptation.

It seems clear that the Hobbesian image of early humankind as leading a life 'solitary, poore, nasty, brutish and short'—an image held by many nineteenth century, and even twentieth century, anthropologists—is not by any means an accurate description of the hunter-gatherer lifestyle. Indeed, as we shall see, there are many reasons to believe that today the average human life shows little improvement over that of our early ancestors. Certainly the hunter-gatherer lifestyle allows a degree of leisure which cannot be equalled in the West except by wealthy people, and their health and longevity almost compare with our own.

Inspite of this, remaining hunter-gatherers are today seriously threatened by the encroachment of agriculture and pastoralism, and this is something that should concern us all deeply. Their disappearance may mean not only the loss of entire tribes with their unique genetic and cultural inheritance, but also the loss of an image of our own past, a past which has done more than any other historic or prehistoric phase to make us the kind of beings that we are today.

References

1   Lee, R.B. 1968 What Hunters do for a Living, or, How to Make Out on Scarce Resources. In *Man the Hunter* R.B. Lee and I. DeVore, eds. (Chicago: Aldine, 30–48).
2   Lee, R.B. and I. DeVore (eds.) 1976 *Kalahari Hunter-Gatherers: Studies of the !Kung San and their Neighbours* (Cambridge: Harvard University Press).
3   Tobias, P.V. (ed.) 1978 *The Bushmen: San Hunters and Herders of Southern Africa* (Cape Town: Human & Rousseau).
4   Jewell, P.A. 1980 Ecology and Management of Game Animals and Domestic Livestock in African Savannas. In *Human Ecology and Savanna Environments*, D.R. Harris, ed. (London: Academic Press, 353–81).
5   Oliver, S.C. 1962 Ecology and Cultural Continuity as contributing factors in the Social Organization of the Plains Indians. *Univ. Calif. Publ. in American Archaeology and Ethnology* 48 (1): 1–90.

8 Pastoralism

'Woe be to the shepherds of Israel ... seemeth it a small thing unto you to have eaten up the good pasture, but ye must tread down with your feet the residue of your pastures?'

Ezekiel 32:2, 18

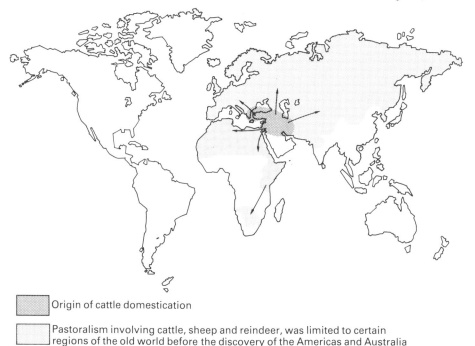

▨ Origin of cattle domestication

☐ Pastoralism involving cattle, sheep and reindeer, was limited to certain regions of the old world before the discovery of the Americas and Australia

Fig 8.1 Pastoralism is believed to have originated in the Middle East, but it does appear to have spread very rapidly. The exact area of origin is therefore still somewhat uncertain.

CHANNELLED PRODUCTIVITY

In the preceding chapters we have discussed a selection of human adaptations to different biomes and we have attempted to assess the place of humankind in the different ecosystems that we have considered. In most of these examples, humankind has been seen to be a component of a stable system, and has appeared to live well within the limit set by the food supply. A balanced ecosystem is, of course, never static, but is always adjusting to changes in one or more components and in their inter-relationships. The system owes its survival to this functional dynamism, which allows it to maintain equilibrium in the face of climatic and other change.

Undisturbed ecosystems are characterized by a typical spectrum of plant and animal life which takes a totally different form in each biome and which can be shown to be the state of equilibrium that is reached under a particular set of climatic conditions. When an ecosystem evolves towards such a 'climax' it gains an increasing diversity of species, and the stability of the final state is a reflection of the diversity that has been achieved. Diversity and stability seem to be correlated characteristics, so that small changes occurring in components of a diverse ecosystem are less likely to alter it in a fundamental way.

As we have seen, the appearance of humans as hunter-gatherers in an existing ecosystem is a factor to which an ecosystem can usually adjust. The humans tend to tap the ecosystem's productivity through a diverse diet and they interact with the system in a highly complex manner, as do the thousands of other components that constitute it. It is only in geologically recent time, since the appearance of agriculture and animal husbandry, that humankind has altered and simplified existing ecosystems in a basic way. With this development in cultural evolution our effect on the environment has been fundamental, extensive, and often destructive.

Farming is the controlled breeding and protection of selected food plants and animals at the expense of wild forms. The overall effect is a reduction in the diversity index: in modern agricultural practice this may eventually lead to the cultivation of only one species over wide tracts of land (monoculture). It is clear that if the energy of an ecosystem can be channelled solely through food plants and animals, we are in a position to tap almost the full productivity of the area which we are farming. The total productivity may be reduced (unless limiting factors are removed), but the amount available for human consumption is usually much increased (unless degradation of the soil sets in). The most modern farming techniques depend not only on the culture of very few species, but on the systematic removal of limiting factors, for instance by the addition of soil nutrients and water. More primitive farming takes place more or less within existing limiting factors. Both types may result in a permanent alteration in the nature of the soil (especially the organic and mineral content) or in its partial or total loss. Such edaphic changes are often irreversible and the resulting wind or water erosion may result in the degradation of the entire system to an unproductive desert.

Soil is one of the components of the world's ecosystems which is, in many areas of the world, a limited and effectively an irreplaceable resource. Soil is the product of the steady effect of climate and vegetation on the rock of the Earth's crust over periods of hundreds and even thousands of years: it can often be destroyed by erosion in a few months. In our review of farming activities, we shall look closely at this effect on the world's soils, for the edaphic factor—often unstable—is the most fundamental component of any ecosystem. In the following pages we shall investigate the place of pastoralism and ranching in human ecology, and their effect on the soil which supports them.

PASTORALISM AND RANCHING

A way of life that appears to lie between that of the hunter and that of the pastoralist, is *transhumance*. We have already seen a version of this among the Tungus of Siberia. It is also found today among the Lapps of northern Scandinavia. Such peoples still follow the herds of migratory reindeer as our own ancestors may have done during the latter parts of the last ice-age. Today there are in this area (Fig 8.2) about 300 000 reindeer and about 30 000 people. As the herds migrate across

Fig 8.2   The Lapps occupy an extensive area of mountainous land in the arctic of northern Scandinavia.

Fig 8.3　The Lapp's relationship to the reindeer is intimate, yet the animals are only partly domesticated. The herd constitutes the main resource for both food and raw materials for Lapp technology. In this photograph a small herd is crossing a glacier in northern Sweden. (*Frank Lane Agency*)

the mountains from one lichen pasture to another, the Lapps go with them, but they do not control the animal's migrations, they merely follow them (Fig 8.3). Amongst these nomadic herders there is only limited domestication: there is no control over the breeding or other important activities of the wild reindeer.

A few, however, are caught and domesticated: some males are castrated and used as draught animals. But it is the wild herd which constitute the main food resource of the Lapps. The people are entirely dependent on the animals for meat and milk, and the by-products such as fur, hides, bones, and antlers. Each Lapp eats one pound of meat per day—this forms the largest part of their diet. Probably the most essential development, however, which characterizes all pastoralists and separates them from hunters, is the use of milk. This is in effect a way of extracting food from animals without killing the animals.

The term transhumance has also been used in a wider sense—to describe a form of pastoralism sometimes called 'nomadism' which involves regular migrations of domestic livestock and their owners

Fig 8.4 Milking of herbivorous mammals represents an extraordinary advance in resource extraction as it produces animal fats and protein for the pastoralists without killing members of the breeding herd. It is an efficient and ingenious approach to the exploitation of mammals. (*Finnish Embassy*)

between sheltered valleys in winter and mountain pastures in summer. Pastoralists of this kind are found in many parts of the world—wherever there are mountains or highland pastures too cold to be grazed except in summer. Clearly this is an extremely efficient way of exploiting the productivity of areas which are uninhabitable and non-productive through part of the year. It probably evolved from the much simpler form of transhumance we have described for the Lapps.

Full pastoralism involves *domestication* of a variety of livestock, from reindeer to llamas. The most important are cattle, camels, sheep, goats, horses, donkeys, and reindeer from the Old World and llamas and alpacas which occur in South America. While pastoralism has been well established for millennia across the arid regions of the Old World, from Manchuria to Morocco, there was no independent development of it in the New World: it was introduced by Europeans[1].

Under the circumstances of domestication, where cattle are protected against predation, they are in a strongly competitive position with the wild herbivores which share their habitat. So long as the domesticated herds are regularly culled, so that their numbers are limited in this way as well as by disease, the herds and their owners have not disturbed in any fundamental way the balance between producers and consumers. Thus, although it is channelled through the human population the energy flow through the ecosystem has not been seriously altered, although the diversity of fauna is reduced.

THE MASAI

In tropical and temperate grasslands the primary production of grass is usually limited, as we have seen, by rainfall and temperature variations respectively, and an annual migration is a typical feature of the behaviour of the wild herbivores and their predatory carnivores. Such small scale migrations (which can hardly be called transhumance) are also practised by pastoralists, and in tropical grasslands the limiting point in seasonal production can usually be recognized at the end of the dry season when game and cattle converge on a few permanent water-holes or rivers.

One of the most informative and best studied examples of pastoral nomadism and its ecological effects is that of the Masai of East Africa. These people are tall slim black Africans who live in the low rainfall savanna grasslands of Kenya and Tanzania, near Olduvai Gorge, and close to the area occupied by the Hazda (Chapter 3) (Fig 8.6). The rainfall here is between ten and twenty inches per annum, but is limited to two seasons. Two dry seasons, one of which is five months long, seriously limit the productivity of the herbage, and thus the carrying capacity of the savanna. The Masai and their herds traditionally follow the seasonal migrations of the game, spreading out on the wide plains during the wet season, and gathering at the water-holes at the end of the dry periods. This seasonal migration means that the animals range over an extremely large area of savanna. The Masai occupy an area of some 90000 square kilometres at a density of about 1.3 per square kilometre (Fig 8.6)*.

---

* The 1960 Census revealed about 117000 Masai, 970000 cattle, 660000 sheep and goats, and an unknown number of donkeys. The effective stocking rate is 1 animal unit per 2.3 hectares. An animal unit is 1 cow or horse, or 5 sheep or goats.

Fig 8.5 The Masai of East Africa have taken pastoralism to its most efficient level. Not only do these people run herds of a number of different species (which between them most effectively exploit the savanna vegetation), but they draw blood as well as milk from their animals. In this way they can maintain and even increase the size of their herds, while they exploit their food value in a remarkable way. (*De Vore, Anthro-Photo*)

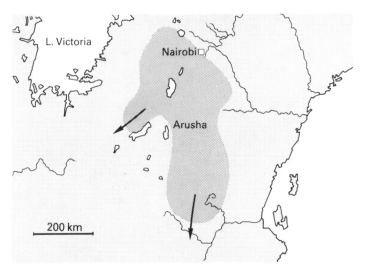

Fig 8.6   The Masai are believed to have expanded southward from the Sudan and Northwestern parts of Kenya in relatively recent times. They compete directly with the rich and varied herbivore fauna of the savanna, as well as destabilising the ecosystem. As a result the survival of the game is seriously threatened.

The domestic livestock carrying capacity of the savanna seems to be roughly similar to the figures we have for the large wild mammal biomass, ranging from 11 000 kg/km² in wetter regions to 3800 kg/km² in the dryer regions. (In more arid areas on the edge of the Sahara, for example, the carrying capacity may be only 1500 kg/km².) To put it another way the capacity varies from 1 animal unit per 2 hectares at 1000 mm rainfall to 1 animal unit per 20 hectares at 200 mm rainfall. (A well-run English farm can carry as much as 2 animal units per hectare, and more with plant nutrients added to the soil.)

The Masai have adapted to their dry savanna environment in a remarkable way*. Their method of extracting food from the herd has been basically modified from that of the hunter: as well as killing, which is usually confined to ritual occasions and special celebrations only, they not only milk the cattle but also draw their blood through a vein in the neck. In addition they collect wild fruit and vegetables when these are in season. In practice they can live almost entirely on animal products and local plants, which together make up their customary diet. Today they buy maize, but they are not dependent on supplies of

* A proportion of the Masai people who live in the wetter regions have taken up agriculture: in the following description, we are considering those who still practise full-time pastoralism.

imported vegetable foods even in the very dry areas in which some of the tribes live. Like the San, they are a well-nourished people and their children show no evidence of protein deficiency.

By switching from dependence on meat to a basic diet of milk and blood, the Masai have developed an economy for the domestic herd which is quite different from that which we practise. In the absence of regular culling, such as is traditional in Europe, the growth potential of the herd is much increased. As a corollary of this, the herds of the Masai become an important sign of the wealth and prestige of their owners: as a result herd size is increased in every conceivable way. Without the private ownership of land—it is all in the public domain— there is no obvious limitation to the size of herds. In a certain district of Kenya (Narok), a 36% increase in cattle was recorded over four years (330000 in 1956; 450000 in 1960). At present, there are about eight cows for each Masai.

It is clear that the potential for increase in numbers of cattle and the desire on the part of the Masai for large herds (regardless of quality) will have certain fairly obvious consequences[2]. In the first place, so long as the human population depends almost entirely on the cattle for its food supply, the potential for increase of population among the Masai will also be maximized. In the second place these numerous large herds of cattle will become liable to suffer epidemic diseases, especially when they crowd at the water-holes in the dry season. Third, the Masai/cattle complex will compete very effectively with the wild game in the area and will tend to replace it. The increase in population of cattle and people will result in a need for more grazing, at the expense of the game. The pressure on the grasslands results in overgrazing and degra- dation, and the pressure for expansion is great. The Masai came from Northern Kenya and have entered their present area only in the last 200 years. They are fearless warriors, and in the absence of the existing political control over their warlike activities they would by now have spread further than they have.

It follows from these considerations that any reduction in numbers of cattle will be bound to affect the Masai adversely and immediately. A serious drought in 1961 resulted in very poor grass production: between 300000 and 400000 cattle died of hunger, together with about 1000 Masai, although water itself was available throughout the year.

Disease has been one of the factors which has been most powerful in the past in limiting the numbers of the Masai and their cattle. Rinder- pest, as well as other diseases (such as foot and mouth disease, pleuropneumonia, and trypanosomiasis), has wrought havoc with their cattle, and smallpox has killed off the Masai themselves in large numbers. Today medical science can control most of these epidemics so that a Masai population of 45 000 in 1890 has risen to 117 000 in 1960 and is still increasing. Their density is now six times as great as it is believed to have been 100 years ago.

The effect of this rapid increase in numbers in what is now a limited area is that the savanna is overgrazed. Having decimated the game by displacement, the Masai herds are now destroying the grass. One of the easiest ways to distinguish a herd of Masai cattle from a herd of wild game is that the cattle graze closely packed together. This alone puts a much greater stress on the grass plants than the broadly dispersed herds of wild animals which graze alongside them and graze selectively.

The destructiveness of the Masai herds has been further enhanced by the activities of the government: in certain areas of East Masailand permanent water supplies were established for the Masai and their cattle. The water was obtained by drilling and by piping spring water from the mountains. This 'improvement' has had two serious consequences: one is that the supplies of water for wild animals have been reduced, and the game has suffered as a result (only one quarter of the original natural water points are still open). The second is even more serious: the perennial water supplies have allowed the Masai to reduce their annual migrations so that grazing is concentrated throughout the year in certain limited areas. In these areas the shortage of grass is acute, yet as long as the Masai can get their cattle to water every other day, they seem to be satisfied, although, as we have seen, the cattle may nevertheless die of starvation. The critical factor is the increasing distance between water and grass.

In the areas where these good water supplies have been introduced the original grass has now deteriorated completely and it has been replaced by thorn scrub and trees: between the trees the soil is bare. An erosion pan of blowing sand and dust—a thorn bush wasteland—is the result. This degradation is not limited in area, but now threatens a considerable proportion of Masailand. The Masai have two short-term solutions which enable them to face this problem, but which in the long run exacerbate it still further: both are examples of positive feedback.

The Masai response is to burn the dead grasses in the dry season so that the young shrubs are killed and fresh young grass is easy to graze when the rains come. They regularly burn the savanna, sometimes as often as four times per year. This does have the advantage of maintaining the grasslands against the encroachment of trees and shrubs, and it seems probable that in many areas the Masai have been able to establish grassland in the place of woodland. (Grass can stand regular burning, while young trees and shrubs cannot.) However, the effectiveness of burning depends on there being enough dead grass on the plains to produce sufficient heat to kill the woody plants. If the savanna is grazed too heavily, even burning will not protect it against the encroachment of unpalatable shrubs and trees. The less grass there is, the more the Masai need to burn it, but if there is less grass, the burning is less effective. Again we see the ecosystem in a process of positive feedback towards degradation.

The increasing emphasis on sheep, goats, and donkeys is another calamity. These small animals can flourish where cattle will starve, because their food requirements are different: they can eat grasses which are unpalatable to cattle and various shrubs and non-graminiferous herbs. This has the result that the total ground cover, which holds the soil in place in the face of heavy tropical rains, is further reduced. Even limited grazing by this combination of domestic animals is destructive: huge erosion pans are the result and the grasslands are becoming desert. The areas most seriously damaged are in Eastern Masailand where the government has introduced water supplies. In the more remote West Masailand, the savanna is in better condition.

This very short survey shows clearly that when the process of positive feedback is triggered in an ecosystem, degradation will set in. As the numbers of Masai herds and people increase, their basic subsistence, the wild grass of the savannas, is being systematically destroyed. We can see that here, as in so many other parts of the world, the veterinary and medical services have lowered the death rate by removing the natural checks on the population expansion of cattle and humans. The extractive efficiency of the Masai people in their dry habitat is so high that they can tap almost the entire production of the savannas. Overgrazing and soil erosion is the inevitable result. If no change is made in the social behaviour of these people, they will ultimately destroy the soil, their habitat, and the last of the wild animals which share it with them. The consequence would appear to be starvation or civil warfare. The solution must involve a limitation in the total number of herbivorous animals in a given area: ultimate loss of the topsoil makes the land unable to support either animals or people.

Fig 8.7   The western ranges of the USA were once grazed by vast herds of buffalo. Being in the public domain, the land has been devastated by overgrazing and the buffalo are almost extinct. That part in private ownership has been much better preserved. In this photograph the topsoil has been washed onto the road in the foreground. (*USDA—Soil Conservation Service*)

GRAZING THE COMMONS

It is easy to account for this disastrous state of affairs by pointing at the ignorance of the Masai people. Yet the degradation is in fact a product of their ignorance combined with that of the Europeans. Before Europeans came to East Africa, the Masai maintained a far smaller population, and degradation was nowhere nearly so extensive. As we shall see, the prospects before the most advanced agricultural areas in the world are not much better. Wherever extensive cattle ranching is carried on in the semi-arid regions of the world, whether in Australia, South Africa, or the USA, there has been a degradation of the grasses and the soil, and the threat or the reality of soil erosion. Mountain lands in the west of the USA were seriously overgrazed in the latter part of the last century and the early years of this one, resulting in depletion and change in the plant cover, and erosion of the soil (Fig 8.7). Although the rate of deterioration has been reduced, it still continues[3].

It has been estimated that the original capacity of the western range of the USA (some 325 million hectares) was 22·5 million animal units. During the sixty years to 1930, the carrying capacity was reduced to 10·8 million units, although the actual number carried by the range was 17·3 million units. It was, and still is, seriously overgrazed. The result is that as early as 1930, 263 million hectares of the total area were suffering from serious erosion, and only 38 million hectares were in a satisfactory condition. This good land is either National Forest or in private hands: the damaged acreage is in the public domain. In 1975, 71% of public land was either static or declining in condition and 83% of the range administered by the Bureau of Land Management was in an unsatisfactory or worse condition. The problem which arises from grazing public land is a worldwide phenomenon: we find it, not only in the USA and East Africa, but in all the other grassland regions of the Old and New Worlds. Serious overgrazing and devastation is recorded in India, throughout the Middle East, and in the Mediterranean region. The Sahara desert was to some extent due to human activity within the last 5000 years, and is presently being extended. In Australia, the combination of sheep and the European rabbit has caused extension of the Great Central Desert.

The problems of grazing the semi-arid grasslands of the world are acute. One of the most promising solutions would be to replace the domestic herds by wild game[4-6]. The plains herbivores are finely adapted to their environment and can live in equilibrium with the delicately balanced ecosystems we have discussed. They can produce more meat per acre, when properly culled, than can be obtained from domestic livestock in the same areas. Cattle could then be reserved for use in the temperate woodland regions, such as we find in much of Europe, where forest clearance and well distributed rainfall allow the maintenance of productive grassland in the face of heavy grazing. In these areas wild cattle originated and cattle rearing gives good economic returns. In place of a density of about one animal unit to 19 hectares which we find in the USA western range, English pastures can easily carry one unit per half hectare. Where they are not irrevocably destroyed the western ranges are probably better suited to carrying bison.

SUMMARY

Domestication can be defined as the capture and taming of animals of

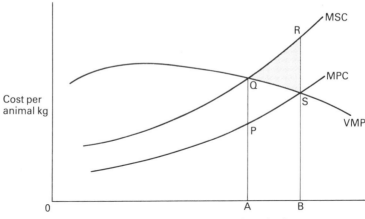

Fig 8.8   The cost of adding animals to a herd grazing a common pasture is shown here. As the number of animals is added to the commons (bottom line) the costs to the individual owner increase a little (MPC) (after the initial stocking) but not as much as the cost to the society (MSC). As these costs increase, the production per animal falls (VMP). Because the cost per animal effects the society (Q) before it effects the individual (S), the individual herdsman will continue to stock the pasture as the social cost grows to intolerable levels (R). (After Hardin[7])

a species with appropriate behavioural characteristics, their removal from their living area and breeding community, and their maintenance under controlled breeding conditions which make possible artificial selection of character traits. The domestication of wild animals was one of the most important developments in human history. It made possible an end to hunting as a way of life and a great increase in the human population.

The concept of territorially defended grazing lands must have arisen early in the process of domestication; but within a single community land was not privately owned among the early pastoralists. The normal state of public ownership of grazing land or common land has at some stage in history had a serious effect upon the ecology of all the areas where pastoralism has flourished. Garrett Hardin has drawn attention to this question in his most important essay entitled 'The Tragedy of the Commons'[7]. Hardin pointed out that in a situation where a number of pastoralists kept their flocks on a limited area of common land, it would pay each flock owner to increase his flock as much as possible and that the advantages to him of an increase of one animal would far outweigh the effect on him of the relatively slight degradation to the pasture caused by that animal. The logic is therefore remorseless: each owner will maximize his herd to the greatest extent and no one individual will suffer in proportion to his increased advantage (Fig. 8.8) Eventually, however, the population will face ruin and the whole social structure of people and animals will crash. The increase of herds beyond the grazing capacity of common land is a worldwide phenomenon: we have described it among the Masai and on the western ranges of the USA. In some western states the ranchers are actually trying to expel the agents of the Bureau of Land Management for their interference with their freedom to stock the ranges as they would wish. The remorseless drive for personal gain is irresistible in spite of Federal Legislation.

In England, overstocking by sheep was equally serious until the Enclosure Acts (between 1709 and 1845) when the powerful landowners took the common lands into their estates as private freehold property. Similar changes in land ownership took place in other western European countries between the sixteenth and nineteenth centuries. The suffering of the shepherds and farmers was great as a result of this development. In retrospect, however, we can see that without bringing in very strict state control over all aspects of pastoralism, private ownership of land is the only way to control productivity and save the integrity of the pastures.

In summary, the commons are practical only under conditions of low population density. As the population has increased the commons must be abandoned. In many places this has occurred, but by no means everywhere. Beside grazing land, rivers, coastlines, the vast majority of the oceans and the air space over the Earth, are all held as common property, in part at least, to this day.

Every new enclosure means the infringement of somebody's personal liberty. 'Rights' and 'freedoms' are lost, but in the end freedom can only be secured by the recognition of essential controls. Pastoralism generated huge food resources for humankind, and brought in its wake far-reaching social developments of extraordinary complexity and importance.

Overgrazing, the tragedy of pastoralism, has brought about chronic deterioration in the ecological communities of many areas of the world. Plant diversity has been greatly reduced to unstable and non-productive levels, wild animals have been displaced and are often almost or completely extinct. Worst of all, soils have been destroyed and grasslands reduced to stony, eroded wastelands, and in some places even to drifting sands. Overgrazing in arid or semi-arid regions of the world has been a disaster for nature and a tragedy for humankind.

References

1   Ucko, P. J. and G. W. Dimbleby 1969 *The Domestication and Exploitation of Plants and Animals* (London: Duckworth, Chicago: Aldine).
2   Talbot, L. M. 1968 The ecological consequences of rangeland development in Masailand, East Africa. Submitted to *Conference on Ecological Aspects of International Development*, Warrenton. (Unpubl.)
3   Anon. 1936 The Western Range, a great but neglected natural resource. *U.S. Forest Service, Senate Document* 199. Washington.
4   Dasmann, R. F. 1964 *African Game Ranching* (New York: Macmillan, London: Pergamon).
5   Talbot, L. M. et al. 1965 The meat production potentail of wild animals in Africa. *Commonwealth Bureau of Animal Breeding, Genetics and Technology. Communication No. 16: 42.*
6   Jewell, P. A. 1979 Ecology and Management of Game Animals and Domestic Livestock in African Savannas. In *Human Ecology and Savanna Environments*, D. R. Harris, ed. (London: Academic Press, 353–81).
7   Hardin, G. and J. Baden 1977 *Managing of the Commons* (San Francisco: W. H. Freeman).

9 Agriculture and pollution

'In the sweat of thy face shalt thou eat bread.' Genesis 3:19

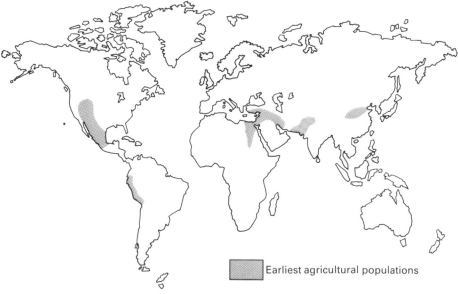

Earliest agricultural populations

Fig 9.1 Agriculture is believed by many scholars to have originated independently as a result of a series of remarkable coincidences in a number of different regions. The earliest was most probably the region of the fertile crescent, the foothills surrounding Mesopotamia. But other developments soon followed in Upper Egypt, Pakistan, China (possibly by diffusion) and soon after in the New World – almost certainly an independent discovery.

HORTICULTURE AND AGRICULTURE

These terms are used to describe two distinct patterns of farming. *Horticulture* is the husbandry of a mixed group of food plants in a garden, near a dwelling. This is a widely distributed means of raising crops, which is most typically found in tropical areas of Central America, Africa, and Southeast Asia. The method maintains a considerable diversity of species within the area cropped, and the gardens tend to be isolated among the surrounding wild vegetation. The great advantage of horticulture is the diversity of species maintained, so that no large stands of single species are planted. This helps to preserve the crops from the sort of epidemic infestation of disease and pest which tend to infest field agriculture. Again we see that diversity carries with it stability, and great oscillations in the productivity of horticultural crops are not usual.

On the boundary between horticulture and agriculture we find the technique of *shifting cultivation* or field-forest rotation which is typically practised in rain forest areas. This method, which is perhaps best called *swidden agriculture*, involves the clearance of small areas of rain forest which are planted with a variety of crops[1]. It is also characterized by great diversity of species, and in the Philippine islands as many as forty varieties of cultivated plant may be grown in a single three acre plot. The high diversity of food plants again places this type of agriculture in sharp contrast to the low diversity of the most common methods of field cropping.

Agriculture or *field cropping* involves growing large areas of one or two species of food plant; more than is needed for the subsistence of a single family. We have already briefly described the agriculture of the Iroquois (Chapter 4) and in this chapter we shall consider advanced examples of agriculture made possible by the development of the plough. This specialized food production is always accompanied by urbanization, which it makes possible. Agriculture limits species diversity in an extreme manner, and this has a number of predictable results. One of the most famous famines of recent times was the result of a dangerous degree of monoculture and too great a dependence on a single food. The introduction of the potato in Ireland in the seventeenth century led to the decline of the great feudal estates. Numerous potato plots nourished larger families and allowed early marriage. The population increased from about 1 million in 1670 to 8·2 million in 1846, and during this period 1·5 million people emigrated. Between 1848 and 1854 potato blight, a disease caused by the fungus *Phytophthora*, spread rapidly over the country and destroyed the crop. The leaves turned black and the potatoes rotted in the ground. Nearly one million people starved to death, and another million emigrated. Since that time the population has stabilized at 4 million as a result of further emigration and a return to late marriage.

In this historic famine, we can see clearly how high agricultural productivity made possible a sharp rise in the human population. But we can also see the danger of monoculture: crowding of animals or plants is an invitation to epidemic disease, and the potato blight destroyed the subsistence food of 8 million Irish. Table 9.1 shows some of the most important crops in the world, in terms of value and production.

Table 9.1   Some of the most important crops in the world

Crops	Production metric tons (millions)	Value US dollars (millions)	Comments
Wheat	343	21000	Mostly human food; good protein
Rice	308	15000	Mostly human food
Maize	308	17000	Much fed to livestock
Potato	306	17000	Mostly human food; high in water
Barley	152	7000	Some fed to livestock
Manioc	92	–	Estimates poor; local consumption
Oats	54	–	Much fed to livestock
Sorghum	49	2500	Some fed to livestock
Soybean	49	6000	Much fed to livestock; high protein and oil content
Cane sugar	41 } 72	6000 } 9000	Low nutritive value
Beet sugar	31	3000	
Citrus	37	–	High in water, vitamins
Cotton fibre	11 } 33	7069 } 8535	Important edible oil
Cottonseed	22	1466	
Bean, pea, chickpea	31	–	High nutritive value
Rye	31	–	Some fed to livestock
Banana	28	2500	Estimates poor; local consumption
Tomato	28	–	High in water, vitamins
Millets	22	1063	Mostly human food, beer
Sesame	21	–	Mostly human food, high oil
Palm oil	20	4000	High caloric value
Peanut	18	3500	High oil and protein
Sweet potato and yams	15	–	Estimates poor
Coffee	4.9	4000	Cash value; caffeine
Tobacco	4.5	4900	Cash value; tars and nicotine
Rubber	3.5	961	Cash value; no food
Cocoa	1.5	937	Cash value and food value
Tea	1.3	1000	Cash value; caffeine

After Harlan (Ch. 10 ref. 2).

AGRICULTURE AND ENERGY

Every species depends for its biological survival and success upon obtaining sufficient energy resources in the form of food. Horticulture and agriculture are clearly further means of channelling and increasing the amount of food energy available to human populations. Thus, for the San bushmen studied by Lee, one hectare produces approximately 690 kcals of wild food per year. In the UK today on the other hand, one hectare of cereals can generate over one million kcals per year[2]. (These and the following data are based on 1968–72 figures.) The difference is astonishing, but it should be noted that the bushmen live in a marginal

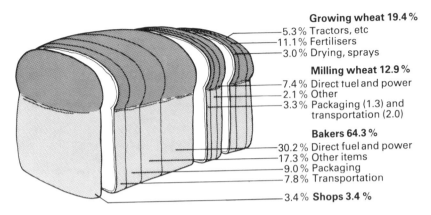

Growing wheat 19.4%
- 5.3% Tractors, etc
- 11.1% Fertilisers
- 3.0% Drying, sprays

Milling wheat 12.9%
- 7.4% Direct fuel and power
- 2.1% Other
- 3.3% Packaging (1.3) and transportation (2.0)

Bakers 64.3%
- 30.2% Direct fuel and power
- 17.3% Other items
- 9.0% Packaging
- 7.8% Transportation

- 3.4% **Shops 3.4%**

Fig 9.2   The energy inputs required to produce packaged bread in England are considerable. For a 1 kg loaf the total input is 5000 kcals. The input from the farmer is about 1000 kcals, from the miller 645 kcals, and from the baker 3200 kcals. The energy value of the loaf to the consumer is 3150 kcals. If the bread were sold fresh at local bakeries rather than packaged and delivered to shops, this would save 20% of the energy input. (After Leach[2])

environment with rainfall well below the optimum, while UK cereal farming is among the most productive in the world. These are therefore extremes of food productivity. Other data are given in Table 9.2. Agriculture, therefore, makes extraordinarily effective use of suitable land for food production.

This enormous output, however, is achieved only at considerable cost, and the cost can be measured in energy as well as money. The energy input to modern agriculture is immense. When we add up all the energy required to cultivate a hectare and to manufacture the machinery, fertilizers, and other chemicals required for such cultivations, then we find that to produce 1000 kcals of output as wheat, we in the UK require inputs of about 620 kcal. (To produce 1000 kcals of white, sliced, wrapped loaf, the inputs rise to 1900 kcals, Fig 9.2.) These figures omit energy input due to farm labour so that the equivalent total inputs for hunter-gatherers are close to zero. Inputs per hectare for some less developed types of agriculture are also shown in Table 9.2.

Thus it appears that from the point of view of an energy budget, modern agriculture is extremely inefficient. However, when we come to consideration of labour inputs, the situation is different: the energy output per man/hour of work among hunter-gatherers is about

Table 9.2   Energy outputs and inputs, in megajoules

Type	Output per man-hour	Output per hectare	Input per hectare	Output: Input ratio
San bushmen*	4.50	2.90	0.37	7.80
Swidden agriculture, Congo*	30.00	15685.00	240.00	65.00
Peasant Farmers, China*	40.00	281000.00	6846.00	41.10
Wheat, UK	3040.00	56200.00	17800.00	3.35
Maize, USA	3800.00	76910.00	29850.00	2.58
Milking herd, UK		10000.00	26900.00	0.37

* includes labour
Note: 4.187 megajoules = 1000 kcalories. The megajoule was used in this table to make the figures more manageable.
After Leach[2].

1070 kcals—among UK cereal farmers it is 726 000 kcals—over 700 times as much. However, we need to remember that a vast number of people are involved in the agriculture and food industry, making tractors and fertilizers, distributing them together with fuel oil, and distributing the food and processing it. When we consider the whole complex of western food production we get a different picture and find that the output per person for UK cereals (for example) is no more than 8360 kcals per hour—about 7·7 times that of the bushmen. In the UK about 5·4% of the whole population (12·7% of the country's labour force) is involved in food production and related industries. Among the hunter-gatherers almost all adults and most children are concerned with either hunting or gathering. The UK produces about 60% of its total food–energy requirement: the rest is imported.

What we have seen in the development of the agricultural process is a means of increasing outputs of food energy at *low* efficiency in relation to energy inputs, but at *high* efficiency in relation to land use and manpower. This seems a logical development, given that energy is relatively cheap and land suitable for agriculture finite in quantity and becoming scarce. By releasing a high percentage of the total population from food production, the division of labour and the development of civilization were made possible.

There are, however, other limiting factors to agricultural development which require consideration beside the very high cost of the energy inputs required for cultivation and fertilizer. Probably water is the most important limiting factor in many parts of the world and

□ Approximate extent of irrigated farmlands

Fig 9.3   The central San Joaquin valley of California is a vast arid plain between the coastal mountains and the Sierra Nevada. With the extraordinarily expensive and extensive irrigation schemes now completed for the entire area, the valley has become the most productive region of its size in the world. It seems unlikely, however, that this level of productivity can be maintained indefinitely without further immense injections of capital, since the ecosystem of the valley has been made unstable.

attempts to overcome this are an important part of the story of human ecology.

IRRIGATION

To understand more clearly the agricultural ecosystem and the importance of the soil, we shall examine in some detail an area which has been brought under intensive production by irrigation. Arid regions of the tropics and sub-tropics have very slight cloud cover, and hence great amounts of sunlight. As a result they have a great productive potential if the limiting factor of water shortage can be overcome. This has been done since early in the history of agriculture, and one of the most famous examples was the irrigation system based on controlling the waters of the Tigris and Euphrates Rivers in ancient Mesopotamia, now Iraq. This system, which may first have been developed as much as 6000 years ago, is only one of a number of early irrigation systems which brought about the growth of extraordinarily productive agricultural regions. The most well-known included the Nile Delta; along the Indus basin in India, and the Yellow River in China: all regions giving rise to a series of remarkably early civilizations. In the New World important irrigation systems were developed early, in the southwest of North America, in Central Mexico, in the coastal valleys of Peru, and in the Incan Andes. Today, China has the greatest area of irrigated land in the world, while in Egypt virtually all farmland is irrigated. Arid regions which have enough river water for irrigation have in some areas become the most productive regions in the world, and in this respect even surpass the tropical rain forest.

One of the most advanced of such agricultural areas is the Central Valley of California (Fig 9.3). About 700 km in length, the valley which lies to the north of Los Angeles is drained by two rivers into San Francisco Bay, the Sacramento in the north and the San Joaquin in the south. The floor of the valley is fairly flat, being made up of alluvial fans formed by the numerous tributaries which drain the Sierra Nevada mountains. In the valley, which is protected from the influence of the cool sea by the coastal mountain range, the climate is hot, and the temperature never falls below freezing. Temperatures of 38°C are frequent everywhere in the summer. The winter rainfall varies from 500 mm per annum in the northern part to 150 mm per annum in the drier south.

This semi-arid basin was originally ranched by the Spaniards, who introduced cattle and sheep, and after 1865, when settlement began, wheat and barley were grown extensively. Corn, beans, citrus, olives, and vines were also introduced, but production was limited, especially in the South, until in the 1880's the first irrigation canals were built. This led to a more diverse agriculture based on fruit and vegetable crops and a vast increase in productivity. By 1930 a plan had been drawn up for the irrigation of the entire valley from the waters of the Sierra Nevada mountains, and this plan is today complete. Water from the Sacramento River is diverted into irrigation channels and carried south to the San Joaquin part of the valley, for this southern flood plain is larger, while its own river is considerably smaller. Thus the natural water supplies of the southern valley are supplemented by water from the much wetter northern parts of California. It is generally believed that well-designed engineering projects to 'save' and transport water will permit continued prosperity in California. However, the fundamental kind of alteration in the entire ecosystem which such extensive irrigation implies could possibly result in an ultimate degradation of the environment.

In the first place, the soils of an arid region such as the San Joaquin Valley often carry a high natural content of salts and alkali (in particular the cations sodium, calcium, magnesium, and the anions chloride and sulphate). Owing to the low and very seasonal rainfall, these salts, which are derived from the weathering of primary geologic materials, accumulate in the surface soils and raise the level of salt constituents beyond the point at which food crops can flourish (about 0·1% by weight). In order to bring these soils into agricultural use, therefore, large amounts of irrigation water are needed to leach out the accumulated salts. This process is effective only if the land is well drained, the subsoil permeable, and the underground water table well below the surface. The absence of proper drainage can cause flooding and the production of permanent salty lakes which will in turn raise the level of the surrounding ground water table. A problem of this kind has arisen in the Imperial Valley of Southern California where the Salton Sea has formed in a depression well below sea level. Here irrigated lands drain into a salt lake and a rise in the level of the lake raises the underground water table and its associated salts in the surrounding region.

Not all irrigated lands are naturally salty. Irrigation can itself, however, bring about the salinization of good soil, since the irrigation water may contain from 0·1 to as much as 5 tonnes of salt per acre/foot of water, and the annual application of water may be as much as 5 acre/feet or more*. Because of this water must be added at a faster rate than that required merely for plant growth and transpiration, or the soluble salts will build up in the surface soils. Since water must be applied in excess of the requirements of the crop, the ground water table may rise, as we have seen, if the land is not well enough drained. Without adequate drainage, soluble salts are added very rapidly to the ground water, and if it enters the root zone it will bring about salinization of the topsoil. It is clear then that the success of the California Water Plan depends on a sufficient and reliable water supply coupled with adequate drainage. If these conditions are economically fulfilled (and this has been possible in many parts of the valley), the agricultural potential of the region could be realized, although the ground water table and the soil salinity must always be carefully monitored[3]. In fact, however,

---

* One acre/foot is the amount of water required to cover one acre to a depth of one foot. It equals 12·34 hectare/cms.

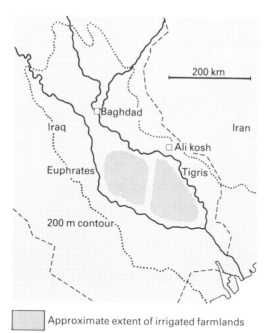

Approximate extent of irrigated farmlands

Fig 9.4   Mesopotamia, the area between the Tigris and Euphrates rivers in Iraq, was once a highly productive agricultural area under irrigation. When the ancient empire and government collapsed the central control of the irrigation canals ceased. In time the system broke down, and today much of the region formally farmed has become desert. (*Geoslides*)

areas of salinization are already appearing where drainage is in-adequate. The water table has been rising and salt is within a few feet of the surface over 160 000 hectares. For the present this means that farmers can cope by switching to the more salt-tolerant crops such as barley, but they already report crop losses of $32 million per annum. The only long-term solution is better drainage: a vast drainage system has been devised with extensive underground drainage leading into an immense ten-foot deep main concrete drain which would convey the salty water 470 km to the sea at a cost of over $750 million. The ecological by-products of this entire scheme may be extremely serious for the future of the valley.

The immense productivity which the California water plan has en-gendered is also fully dependent upon an adequate water supply and yet that water supply itself is in jeopardy. The mean annual rainfall on the most productive 30 million hectares of California has dropped by one third since 1850 (from 840 mm to 560 mm) and this is believed to be mainly due to alterations in the plant cover of the land, through tree-felling and erosion. The creation of National Forests over much of the Sierra Nevada has halted this destruction in much of the area, but it still continues without abatement elsewhere. The greatest industry in California, the farming industry, depends indirectly on the preservation of the forest.

A failure in control of irrigation and in the maintenance of good drainage may quickly bring about the formation of desertic salt flats such as those around the Salton Sea. This type of degradation has also occurred in a number of other once fertile irrigated regions. One famous example is the Tigris–Euphrates system in Iraq (Fig 9.4). The immense fertility and productivity of this region continued for hundreds of years from the time of the Babylonian Empire until the land was abandoned in the twelfth and thirteenth Centuries. As long as strong governments existed to ensure maintenance of the drainage system and silt clearance from the canals, the country prospered. When the governments failed to enforce dredging the canals, the flow of water was reduced and the level of salinity increased with disastrous results to the fertility of the soil. Remains of civilizations beneath the sands of Iraq show clearly that salinity was responsible for the decline of cities and the movement of populations. The valley may once have supported between 17 and 25 million inhabitants: today the total population of Iraq is only about 9 million.

The most serious recent example of degradation has occurred in the

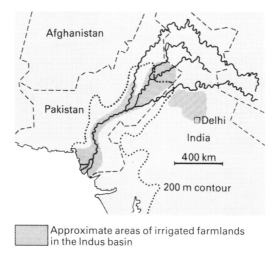

Approximate areas of irrigated farmlands in the Indus basin

Fig 9.5   The Indus basin is another large area which has been developed with extensive irrigation. Here the problem of salination has been temporarily controlled by pumping out the saline water and lowering the water table, but the cost is considerable and an ever continuing burden on the farmers. (*Geoslides*)

Indus basin of Pakistan—an area of over 400 000 square kilometres (Fig 9.5). In this vast plain some 13 million hectares were brought under irrigation by Muslim rulers and later by British engineers to form an area of great productivity. However, poor drainage brought about waterlogging of over 4½ million hectares and the productivity of over 6 million hectares was reduced or destroyed by salinity[4]. With a population increase of nearly one million persons per annum, Pakistan lost one productive acre every five minutes. Having become dependent on the productivity of irrigated land, the failure of some of this land through salinity and waterlogging was a serious development. The problem in Pakistan has been overcome by the development of tube-wells—closed cylindrical shafts which can be driven into the ground—from which ground water can be extracted and pumped into the rivers by diesel or electricity. Thus they are able to lower the water table and at the same time make it possible to increase water use so that the salts are washed out from the soil. The salty waterlogged land is today steadily being reclaimed at a rate somewhat faster than it was being lost.

In the examples of artificially irrigated farm lands we have described, we see that the stability of the ecosystem depends on monitoring by agricultural experts. This means that any disruption of social stability in the area, as well as failure to monitor critical variables, can result in positive feedback and degradation. Once the productivity of the land has declined, the capital to carry out further engineering work is generally not available. Beyond this, however, variables are involved which are beyond the control of the farmers or the department of agriculture. The irrigated ecosystem is not limited by the valley edge— it is a subsystem of a larger system, and is ultimately dependent on factors beyond the control of the people who depend upon it.

AGRICULTURAL POLLUTION

The world increase in population has been made possible by advances in medicine and by increased production of food. The development of agriculture has involved its expansion over vast areas of the Earth's surface so that today almost all suitable land is under cultivation. An increase in productivity has since become possible by removing some of the limiting factors which remain—not only by introducing irrigation schemes, but also by the addition of chemical soil nutrients, and the use of various plant and animal poisons. The use of artificial fertilizers,

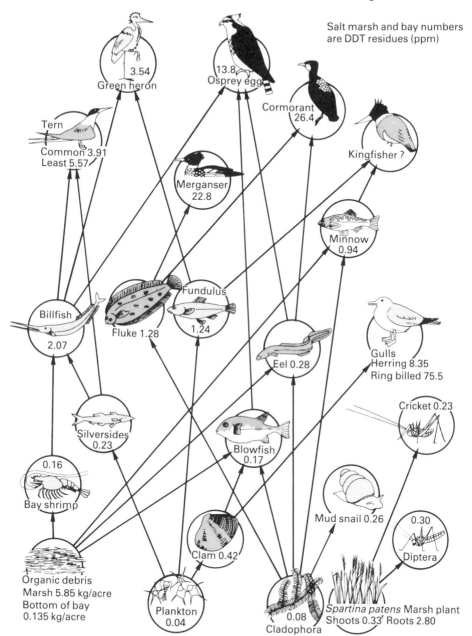

Salt marsh and bay numbers are DDT residues (ppm)

3.54 Green heron

13.8 Osprey egg

Cormorant 26.4

Tern Common 3.91 Least 5.57

Merganser 22.8

Kingfisher ?

Minnow 0.94

Billfish 2.07

Fundulus 1.24

Fluke 1.28

Eel 0.28

Gulls Herring 8.35 Ring billed 75.5

Silversides 0.23

Blowfish 0.17

Cricket 0.23

Bay shrimp 0.16

Mud snail 0.26

0.30 Diptera

Organic debris Marsh 5.85 kg/acre Bottom of bay 0.135 kg/acre

Clam 0.42

Plankton 0.04

0.08 Cladophora

Spartina patens Marsh plant Shoots 0.33 Roots 2.80

Fig 9.6 Increasing concentration (ppm) of DDT in the body fat of living organisms in salt marsh and bay areas. Poison moves up food chains to be accumulated at the higher trophic levels. Thus the creatures most at risk are the carnivores such as ospreys and cormorants.

insecticides, and fungicides, as well as herbicides to control weeds, is rapidly expanding. The existing population has therefore become dependent on agricultural chemicals for its survival, and the projected increase in population will depend even more fully on the products of the chemical industry.

The ecological disturbance associated with agricultural chemicals arises from the fact that the insecticides are powerful poisons which affect a wide spectrum of animal life, while the fungicides and herbicides are poisons which affect a wide range of plant life. The first of the dangers of insecticides became apparent quite early as a result of their effects on farm workers. This was hardly surprising as the insecticides are powerful nerve as well as stomach poisons, and many were developed from poison gases prepared during the Second World War. Less obvious effects are suffered by the consumers of the 'protected' crops. Cumulative poisons have a long-term deleterious effect on mammalian physiology*.

What is of interest to us is the long-term effect of these poisons on the entire ecosystem, which is clearly seen in both terrestrial and marine environments. The effects are serious primarily because many insecticides were developed for their persistence, and some are extremely long-lived under natural conditions. Those that do break down often form equally poisonous and stable substances.

Even if they were not absorbed in fatal doses these stable poisons, such as the chlorinated hydrocarbons DDT and Dieldrin (now no longer widely used), were passed up the terrestrial and marine food chains from producers to consumers, and at each trophic level the poisons became more highly concentrated until they became lethal (Fig 9.6). Their effects were therefore most obviously seen in the carnivores of the second or third trophic levels. Here we find some well-known examples: the Peregrine falcon vanished from the entire Northeast United States as a breeding bird and was at one time threatened throughout England. There were extensive losses in both countries among eagles and hawks, as well as among foxes and badgers. In England 1300 foxes were found dead during the winter of 1959–60 and the cause was traced to the consumption of dead seed-eating birds with

---

* The herbicides, not generally considered poisonous to animals, are not exempt. Research and experience have shown that 2,4,5-T causes birth malformations in rats, and abnormality in new-born children has increased in Vietnam villages where the US used this powerful chemical as a defoliant. Recently, US veterans who fought in Vietnam, have also developed serious symptoms as a result of exposure to 2,4,5-T.

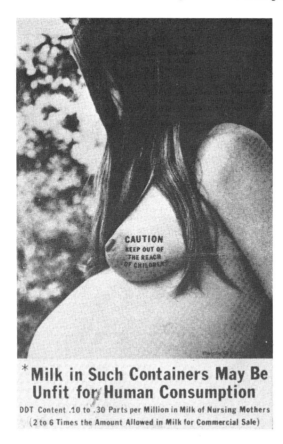

*Milk in Such Containers May Be Unfit for Human Consumption

DDT Content .10 to .30 Parts per Million in Milk of Nursing Mothers
(2 to 6 Times the Amount Allowed in Milk for Commercial Sale)

Fig 9.7 Humans lie at the top of a wide range of food chains, both terrestrial and marine. Accumulation of DDT in human body fat has occasionally caused disease, but reached its most striking manifestation in California mother's milk in the 1960s.

a high residue of Dieldrin in their bodies. The poisons had been applied to seed corn which had been eaten by pigeons, rooks, and pheasants, and a lethal dose could easily be accumulated by a bird which was an efficient seed digger. In Clear Lake, California, a popular fishing resort, DDD (closely related to DDT) was used in 1950 for midge control; a year later the breeding colony of the western grebe (1000 pairs in 1949) was gone. The DDD was at a low concentration of 0·02 ppm (parts per million) in the water, but it had been absorbed by the plankton (5 ppm) and passed to the small fish (10 ppm). In the visceral fat of the predatory fish and the grebes it was found at very high levels of concentration (1600 ppm). Most of the fish caught are eaten by the anglers, and in the visceral fat of the largemouth bass, 1700 ppm were recorded. We now know that sub-lethal doses of chlorinated hydrocarbons can cause sterility, abnormalities in the embryo, failure in egg-shell production, and alterations in behaviour of certain birds.

The passage of chlorinated hydrocarbons up the food chains, first brought to popular attention by Rachel Carson in 1962[5], is an instructive case of ecological shortsightedness. It is now illegal to use these poisons in the USA and most European countries, for we have evidence that deaths from cancers (including leukaemia) and high blood pressure are associated with excessive exposure to chlorinated hydrocarbons. In 1969, some California mothers' milk contained more DDT than that permitted in cows' milk* (Fig 9.7). While our large body size has saved us from accumulating lethal doses, the situation could change if the total level of pesticides in the environment continued to increase. There are at present estimated to be one billion pounds of DDT in the biosphere, and the majority of this will eventually be concentrated at the top of the food chains, much of it in human body fat.

The problem of the chlorinated hydrocarbons is arrested, but it is only part of the problem of agricultural pollution[6-10]. In most fully industrialized countries the use of agricultural chemicals is very extensive, and the food carries high levels of residues. The effect on the soil ecosystem is bound to be serious. The destruction of insect predators will have a very detrimental effect on the entire soil fauna. The presence of these biocides is clearly detectable in the waters of the Sacramento and San Joaquin Rivers which carry a considerable concentration of them, together with a high level of the natural soil salts and the chemical fertilizers. Today, because of the take-off for irrigation and resulting transpiration and evaporation, the outflow of the river system in the delta is less than one tenth of the original outflow. This small amount of water still carries the entire load of salts which were previously carried by the original river system plus the agricultural chemicals. The outflow of this poisoned effluent disturbs the delicately balanced ecosystem of the delta (where the fresh water outflow keeps the sea at bay), and in turn affects the whole of San Francisco Bay and the Northeast Pacific (Fig 9.3).

Eventually much of our poison effluent enters the oceans via our rivers, and here we may find the most striking and dangerous effect of long-term agricultural pollution. A serious reduction of sea birds at the

---

* The permissible amount allowed by the US Food and Drug Administration in cow's milk is 0·05 ppm. The amount in some human milk varied between 0·05 and 0·26 ppm.

top of the marine food chain has already been recorded, and residues of DDT have been found in seals and penguins as far away as Antarctica. But beside this food-chain poison concentration, similar to that we have already described for the terrestrial ecosystem, a direct effect on certain organisms has been observed which may be even more dangerous. Chlorinated hydrocarbons in concentrations as low as a few parts per billion may depress the rate of photosynthesis and cell reproduction in marine algae. As producers, algae are as important in the ocean as grass is in the terrestrial ecosystem. Such chemical interference in algal growth could theoretically prove disastrous to life[11].

In this section we have touched on one or two examples of the way agricultural chemicals may have a fundamental effect on the ecosystem of the entire world. We have seen that the creation of simple artificial ecosystems can itself be a danger, while its effect upon the surroundings upon which these artificial ecosystems depend can be much more serious. The use of agricultural chemicals should be controlled and monitored, and the population explosion which calls for more food and therefore more chemicals should be inhibited by controlling the birth rate. Humankind will not be the first instance of an overcrowded species which suffered from the toxicity of its own waste products, but will be the first to kill at the same time a proportion of the other living species which together make its continued existence possible.

AGRICULTURE, POWER, AND PROPERTY

The extraordinary productivity of modern farming and agri-business is therefore a product of a number of factors, all of which are only made possible by immense capital investment. The California water plan cost $3000 million to complete in the early 1970's, and the new drainage scheme will cost at least another $750 million. Most large irrigation schemes require the kind of investment that only wealthy governments can supply. But agriculture also depends on many other kinds of capital intensive infra-structure including land-levelling (associated with irrigation technology), fencing, road-building, and the construction of granaries and other food storage facilities. More obvious, perhaps, are the chemical and mechanical inputs we have discussed and the factories required to produce them. This industry, which employs thousands of people, requires immense power inputs as well as raw materials, and oil is at present a vital component of both.

Taking into account the extent of these massive inputs, it is obvious that the high productivity of industrialized agriculture is achieved at immense cost and only wealthy societies can maintain it. Food will no doubt continue to be produced in the quantities made possible by industrialization so long as people can afford it. But many raw materials and all fossil fuels are limited in quantity, and as they become scarce, prices will rise. Much of the world's present malnutrition is due to the fact that the people who are starving are unable to pay the high cost of buying shipping, and distributing surplus US cereals.

The switch from a very low input agriculture such as swidden agriculture, or the dry farming practised by the Spaniards in the early days of California (and still practised in many other parts of the world) to the power-hungry modern agri-business has been a change of profound importance. It has enabled humankind to expand its population from perhaps little more than 75 million to its present 4·5 thousand million; but in turn it has put societies and nations in positions of interdependence. The countries producing raw materials are in many cases dependent on industrialized countries for sufficient food, while the latter depend on the former for their essential raw materials. England and the USA have enough domestic oil production to operate their agriculture for many decades, but if some oil is not reserved for this purpose, alternative power sources will soon become vital to both countries. There is no way back to low input agriculture as long as the population does not substantially decrease.

One other aspect of agriculture is of central importance to human cultural development. The mobility of game means that hunter-gatherers do not rely for their survival on certain very specific areas of land, nor do they invest their resources in such areas. This, however, is characteristic of agriculturalists, even the most primitive of whom put great effort and other resources, such as seed, into the plots they are cultivating. Thus their plot represents wealth, and the size of the plot and quality of the soil determines their capital. The worked land is property—property to be jealously guarded and if necessary fought for. The possessions of hunter-gatherers are very few and necessarily portable, and there is for all practical purposes an equality of wealth among individuals. As soon as land can be enclosed and worked, the landowners and the landless will form the rich and the poor of an agricultural community—and successful farming, which depends as we have seen on enclosure and private property, can produce considerable wealth.

The possibility of accumulating wealth is, in a sense, therefore, a product of agriculture, which thus created capitalism. Farmers in productive areas should be able to produce surpluses which can be traded for more land or other goods of lasting value. The production of such food surpluses was an essential prerequisite for the development of another central feature of human cultural development—the city.

References

1   Conklin, H. C. 1969 An Ethnoecological Approach to Shifting Agriculture. In *Environment and Cultural Behavior* A. P. Vayda, ed. (New York: Natural History Press).
2   Leach, G. 1976 *Energy and Food Production* (London: IPC Press).
3   Richards, L. A. 1969 *Saline and Alkali Soils* (Washington: US Department of Agriculture).
4   Wahab, A. 1964 The Soil Problems of West Pakistan. *In The Role of Science in the Development of Natural Resources, with Particular Reference to Pakistan, Iran and Turkey* M. L. Smith, ed. (Oxford: Pergamon).
5   Carson, R. 1962 *Silent Spring* (Boston: Houghton Mifflin, London: Penguin).
6   Randolf, T. G. 1962 *Human Ecology and Susceptibility to the Chemical Environment* (Springfield: Thomas).
7   Rudd, R. L. 1964 *Pesticides and the Living Landscape* (Wisconsin: Wisconsin University Press).
8   Mellanby, K. 1967 *Pesticides and Pollution* (London: Collins).
9   Wodka, S. H. 1970 Pesticides since Silent Spring. *In The Environmental Handbook.* G. de Bell, ed. (New York: Ballantyne).
10 Woodwell, G. M. 1967 Toxic substances and Ecological Cycles. *Scientific American* 216: 24-31.
11 Wurster, C. F. 1968 DDT Reduces Photosynthesis by Marine Phytoplankton. *Science* 159: 1474-1475.

10 The city

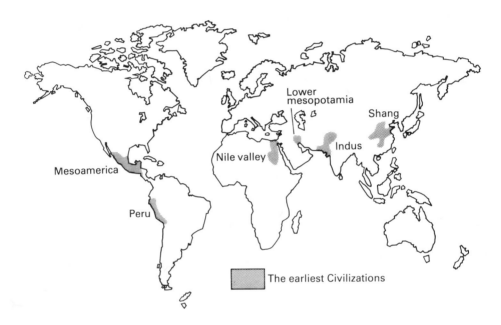

Fig 10.1 Centres of civilization: the earliest cities developed in regions of agricultural wealth.

THE RISE OF CITIES

As we saw in Chapter 6, what seem to be permanent settlements first appear in the archaeological record during Magdelenian times, and are associated with rich riverine food resources. From this time onwards, wherever food resources were available throughout the year in a single locality, relatively permanent settlement could and probably did follow, and there are many examples in the archaeological record, from the early coastal cave dwellings in South Africa, to the first hutted settlements in Siberia. Where food supplies were large and reliable enough, small villages may have grown. The nomadic life of the surviving

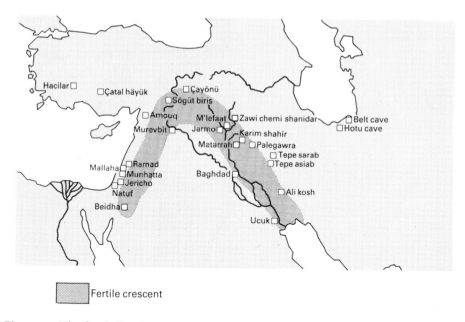

Fig 10.2   The foothills of the mountain ranges surrounding the Tigris-Euphrates basin constitute the fertile crescent. Wild barleys and wheats grow in the mountains of this area, and the foothills made an ideal nursery for the development of agriculture between 15 000 and 10 000 years ago. Later, farming spread to the irrigated plains of the Tigris and Euphrates rivers. Today, rainfall is somewhat reduced, and the area less productive, but the wild mountain cereals still thrive.

hunters and gatherers we have discussed is therefore in part due to the marginally productive areas in which they live. Those in more fertile areas no doubt built permanent or semi-permanent villages, and eventually developed agriculture.

In the fertile crescent (Fig 10.2), where agriculture is believed to have been first developed, it seems very probable that it was preceded by an increasing dependence on wild cereals, which could be harvested and stored just as domesticated varieties are today; wild barleys (*Hordeum* sp.) and wheats (einkorn and emmer of the genus *Triticum*) are still found in abundance throughout this region (Fig 10.3)[1]. Jack Harlan, an American botanist, harvested in the mountains of Southwest Turkey four pounds of wild emmer wheat in an hour, using a Palaeolithic flint-bladed sickle[2]. As soon as harvesting wild cereals yielded surplus food for storage, the immobile food storage containers—often pits— would have encouraged settlement and led to the development of the concept of property, as we have seen. So permanent settlement would have come about gradually, as reliable food collection, and later pro-

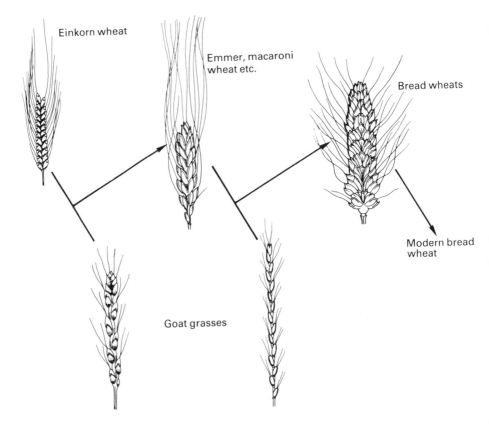

Einkorn wheat

Emmer, macaroni
wheat etc.

Bread wheats

Goat grasses

Modern bread
wheat

Fig 10.3   The wild grains are anatomically distinct from their successors, which
have been selected for thousands of years. (After Heiser[2])

duction, were established. These developments we might have predicted
would be found to occur in areas of high natural fertility and the
archaeological evidence appears to support this.

The most complete (but still relatively sparse) data on early settlement come from the fertile crescent. Archaeological evidence indicates
that by 11 000 BP the first villages had become established (Fig 10.4)[3].
The sites of the villages contain plant remains such as wild einkorn,
emmer, (Fig 10.3) two-row barley , pea, lentil, and vetch. By 8000 BP
agriculture seems to have been established, and we find evidence for
the above plants together with six-row barley, domesticated wheat,
chick-pea, flax, and remains of domestic sheep, goats, cattle, and pigs[4].
The settlements developed and grew on the basis of this broad range of
agricultural produce.

The villages, some of which were to become cities, were therefore

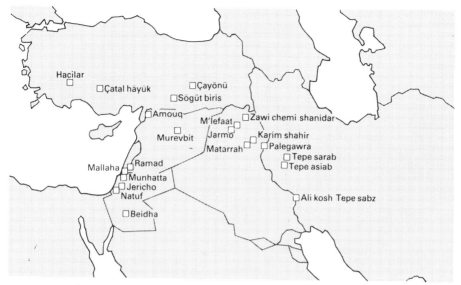

Fig 10.4   Villages developed rapidly around this fertile area and reflected the growth of agricultural production. Population increase followed. These villages date between 9500 and 8500 BP.

totally dependent on the agricultural industry which supported them, and, equally important, on an adequate water supply. The daily availability of fresh water is vital for all higher primates, and human settlements required reliable and pure water in the form of wells or rivers.

As the settlements produced agricultural surplus, trade followed, together with a merchant class. As individuals specialized in crafts, and later arts, merchants began to trade in their products as well as in agricultural surpluses, and trade routes appeared between cities. Those cities which had a flourishing agricultural base and were on the trade routes, at important river crossings, or at the feet of passes, grew and flourished. Thus urban civilization was born.

The cities needed not only food and water, but raw materials for their crafts. Resources such as clay for pottery, wood for carpentry, and ores for smelting metal, became increasingly important and sought after as trade developed. Eventually, settlements which controlled sources of valuable raw materials could afford to buy in food supplies from a wider hinterland and were therefore able to develop in areas of fairly low agricultural productivity. Food, raw materials, and manufactured goods began to be carried from place to place more and more as settlements grew. Transportation first by humans and later by wheeled carts drawn by animals was an important feature of city life. The energy for transportation in the form of draft animals such as oxen, water buffalo, and later horses, became a vital key to urban growth. Thus the progressive domestication of draft animals was profoundly important to this new development.

In temperate regions or high altitude settlements in the sub-tropics, energy was also required for warmth and later for smelting metals. Trees were felled and vast quantities of timber used. Where goats and sheep were kept, the forests were destroyed for ever and deserts were created in the more arid regions of the fertile crescent. The effect on the landscape has been literally devastating and in most places is irreversible. The city's greed for energy was, and is, remorseless.

Thus, as the simple settlement developed into a growing city, it attracted to itself immense quantities of food, of raw materials, minerals, and timber, followed by peat, coal, cattle, dung, etc., and quantities of manufactured products from other places. The energetics of the natural ecosystems were transformed by the complexities of human economic systems. The invention of money further encouraged trade and manufacturing activity. Money—a simple and symbolic means of exchange of goods with a value established by convention—had an almost revolutionary effect on trade. As John Stuart Mill wrote: 'It is a machinery for doing quickly and commodiously what would be done, though less quickly and commodiously, without it.' It is a symbol of value, or energy, and through its convenient exchange, can bring about an immense development in the web of energy exchange in a growing society. Thus, by facilitating trade, money (at an unknown, early date) facilitated many of the most typical characteristics of city life*. The essential basis for this development, however, remained, as it does today, the agricultural fertility and productivity of neighbouring regions.

Trade and concepts of value depended on measurement of quantity, weight, and size. It seems certain that weights and measures must have been developed at this time, together with simple weighing machines, such as balances, and measuring rods. The earliest records we have are from the Middle East—the ancient Egyptians had recognized a standard 'cubit' of 20·62 inches by 5000 BP. It was based on the length of the arm from the elbow to the finger tips. Areas of land were also measured by the early farmers, and weights were probably related to the amount

* Money can take many forms, from cattle or tobacco, to beads or stone discs. Metal coinage appeared about 4000 BP and standardized metal coins about 2700 BP. Paper money did not appear in Europe until 18th Century AD, though it was introduced centuries earlier in China.

that could be carried or hauled by a man or domestic animal. All these developments, that we take for granted, were essential for the growth of patterns of trade and they in turn brought about the invention of simple mathematics.

SPECIALIZATION

The development of the city brought about a proliferation of human activities. Crafts, which had earlier been confined to the preparation of stone tools, of simple leather objects or clothing, and of products made of vegetable fibres or wood, could now multiply to include a whole range of new tools, devices, and facilities. We also find the introduction of ceramic objects such as plates, bowls, and jars, and eventually products in metal. Utensils of all kinds, jewellery, clothing, furniture, and housing were created in many forms for both the poor and the wealthy, as well as for export. Scarce materials and goods of all descriptions were imported by traders and merchants. Eventually certain cities specialized in particular products and became famous for the excellence of their crafts.

As human ingenuity brought about the development of an ever increasing range of manufactured products, craftworkers became increasingly specialized, so that many were able to perform only a limited

Fig 10.5   The flow of energy through a city in the form of food, fuel, and raw materials of every kind is immensely complex, and in this diagram is greatly simplified. The principles, however, are not different to those of a natural food web in nature: we are still part of nature.

range of activities and were dependent upon others not only for their food supply but also for other essentials such as water and fuel. Thus specialization was necessarily accompanied by interdependence, and this interdependence produced an 'ecological' web of goods and energy no different in principle from the food webs we discussed in Chapter 1. Energy in the form of food, or materials, or highly valued goods, passes from hand to hand. It enters the city as heavy and often bulky raw materials: it leaves it as finished products smaller in mass and greater in value (Fig 10.5). The interdependence of the components of this system is as complex and fascinating as the interdependence of the components of a natural food web.

The system operates on a basis of mutual trust and enlightened self interest. Everyone benefits from it as long as the system is well balanced. Imbalance related to food and energy supply can bring about malnutrition or famine and the breakdown of the social system.

Most of the ancient empires can be looked upon as the attempt on the part of a city or state to ensure a constant and reliable food supply. Imperial Rome imported vast amounts of grain annually from Sicily, North Africa, and Egypt. The grain was paid for by taxes and tribute that was owed to the imperial treasury, and it was distributed free to all resident citizens of Rome, possibly as many as 100 000 people. The Aztec capital Tenochtitlan is estimated to have received annually as tribute over 50 000 tons of food together with cloth and metals. As no draft animals were in use in the Americas before contact with Europeans, all these goods were transported by humans. Political power was used to guarantee resources and to extract them at minimal cost.

As rural settlements, following agricultural innovations, increased in size, the country was able to send surplus population to the cities, and still does to this day in most parts of the world. In the past this was usually desirable since the city was often expanding rapidly and the city birth rate was lower than the rural birth rate (and it still is in most parts of the world). A too rapid movement of people to the city, however, has created a class of urban poor that today presents governments with one of their greatest problems (Fig 10.6).

DISEASE

The low city birth rate is a highly significant character of city life, for it reflects the fact that in human settlements where populations became dense, the absence of sanitation allowed the development of a host of

Fig 10.6   City slums seem to be an inevitable product of city growth. They can only be finally eliminated when the population and its distribution between country and city is finally stabilized. This is a typical Central American city slum in El Salvador. (*World Bank Photo, Jamie Martin*)

infections and contagious diseases not very important among rural settlements, including such devastating killers as cholera and plague. Disease can be transmitted in four predominant ways:

1   Airborne droplets from coughs and sneezes (e.g. measles, whooping cough).
2   Faecal contamination of food and water (e.g. typhoid, diarrhoea, cholera, and intestinal parasites).
3   Vectors such as mosquitoes or fleas (e.g. malaria or plague).
4   Contact (e.g. syphilis, scabies, trachoma).

Clearly the first two groups are likely to become a major problem in dense human settlement, while the success of the second two groups of diseases will depend on the adaptability of the vector to city life and the amount of individual contact the culture allows. In Croydon, near London, as recently as 1937, there were 341 cases of typhoid with 43 deaths. The infection was traced to a single workman who had been working on one of the city's wells and proved to be a typhoid carrier. The great plagues of the fourteenth century were due for the most part to bacilli carried by fleas from the millions of rats which inhabited every European city. In England it has been estimated that one-seventh of the population of London died during the Great Plague of 1664–65. The other reason for the spread of disease in cities is the lower nutritional level of many of the citizens—especially the urban poor, who were thus prone to develop diseases—and the early exposure of their children to these infections.

The factor of urban disease was so important in human history that it has been responsible for far-reaching cultural developments relating to cleanliness, eating, greeting, and a host of minor details of behaviour limiting and controlling human contact, which cannot be catalogued here. At the same time some of the earliest cities such as Knossos in Crete and Hanappa, as well as Imperial Rome, went to enormous lengths to build fresh-water aqueducts and tanks, baths, lavatories, and drains and by 600 BC a great sewer, the *cloaca maxima* was being constructed in Rome. The transportation and disposal of waste products became as important as that of food: the natural energy cycle of the human population had to be adjusted and in some cases profoundly modified in adaptation to the density and size of the city: pests and disease organisms had somehow to be controlled, even though the existence of micro-organisms was not yet recognized.

TECHNOLOGY AND CULTURE

As we have seen, technology in its most easily recognizable form as stone tools, is certainly over 2·0 million years of age (Chapter 3), and in the form of simple bone and wood tools, may well be very much older.

The use of tools has always been primarily (if not totally) concerned with the extraction of food resources from the environment. Sticks were used to dig roots, containers to hold small fruits, snails, or insects, and sharp stone flakes to skin animals and cut meat. Heavier hammer stones were used to smash and crush bone to extract marrow. These tools made it possible to increase the food extraction capability of the species in an impressive way. The further development of technology was similar in function, and brought about an even more effective extraction of resources.

The technology of facilities (Chapter 4) was also an important development in prehistory and it made possible adaptation to cool zones otherwise unsuitable for humans. In this way further resource extraction was made possible and the biomass of the species greatly increased. From a much later date machines were in use, first in the form of weapons, such as the bow and arrow and the blow gun, and these too increased the extractive potential of those populations which were dependent upon hunting. Thus the three first subdivisions of technology —tools, facilities, and machines—were developed to enable humans to increase their food supply and their numbers. By the tenth millennium BP, following an extremely slow but accelerating development, humankind's primitive technology was surprisingly advanced. The further development of basic technology increased rapidly with the coming of cities, and advances in agricultural techniques, especially irrigation, made possible rapid city growth.

The city brought people together in large groups and ideas could be exchanged widely as they had never been before. While we must never underrate the technology of hunters and gatherers which was almost certainly more complex than most people suppose, there is no doubt that a further and much greater development in technology occurred in the cities.

Other kinds of creative activity were also becoming important. The arts developed and flourished: the earliest cave paintings (24 000 BP) in Southwest France are associated with more or less permanent settlements in cave mouths and rock shelters. The cities brought a great acceleration of artistic expression and artistic decoration began to appear in most areas of craftwork. Luxury goods beautifully designed and made were sold for much higher prices than utilitarian products. Beauty was appreciated for its own sake and valued: as a result many artefacts became artistically embellished—one of the earliest examples of what looks like non-functional craftwork (Fig 10.7). However, because the more work put into a product the greater its value, artistic embellishment did serve the function of generating wealth without increasing raw material inputs. Where surplus labour is available wealth can be created by up-grading products in this way. A classic example is the production of the finest Persian rugs in rural villages. Art served both industry and religion.

One other aspect of technology became important later—the development of *instruments* such as clocks, microscopes, etc. which made it possible to find out more about the environment and the Universe than was previously known. It became clear that to increase our resource extraction rate we needed to understand natural processes and

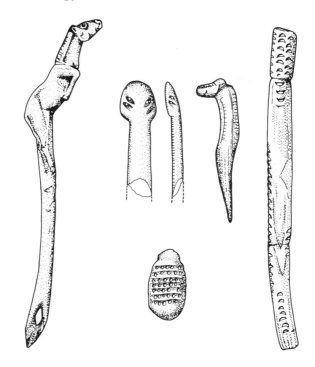

Fig 10.7   The earliest cave art in Southwest France and Northern Spain is
associated with some degree of permanent settlement and of Magdalenian age. It
was, however, fixed in one place and created for ritual purposes, rather than
trade. Soon after this, small objects were decorated and exchanged, such as these
Magdalenian bone tools. .

substances in more detail. The observation and understanding of
Nature—the Natural Sciences—eventually were to become the main
keys to increasing resources.

Thus we can see that humankind's technological equipment constitutes the species' unique adaptation—its way of increasing resource
extraction to a level of effectiveness far above that seen in any other
species.

Culture is, however, more than technology. To paraphrase our dis-
cussion in Chapter 1, culture can be defined as the totality of what a
human society practises, produces, and thinks, that it is possible to
transmit by learning. Culture is the non-biological part of society's
adaptation to its environment, and therefore it is not surprising that
cultures vary as much as environments do.

The most obvious ecological characteristic of cities is that they
present their human inhabitants with a completely new and altogether

different environment to anything else found on Earth. They constitute as distinct an ecosystem as an entirely new biome. They have therefore evoked a new and special kind of human cultural adaptation including a new technology. The outstanding social novelty is the highly developed hierarchical organization so characteristic of large highly organized social groups; the outstanding technological innovation is perhaps in the field of transportation, and centred about the invention of the wheel. Until the appearance of cities the transportation of goods (or people) was of almost no importance to human populations. Today city life depends for its very existence on an effective and very complex transportation system. Thus the city environment has generated many of humankind's most typical and widespread cultural and technological achievements.

References

1   Heiser, C.B. 1973 *Seed to Civilization: the Story of Man's Food* (San Francisco: W.H. Freeman).
2   Harlan, J.R. 1975 *Crops and Man* (Madison: American Society of Agronomy).
3   Ucko, P.J., R. Tringham and G.W. Dimbleby 1972 *Man, Settlement and Urbanism* (London: Duckworth).
4   Ucko, P.J. and G.W. Dimbleby 1969 *The Domestication and Exploitation of Plants and Animals* (London: Duckworth).

11 The human ecosystem: past, present, and future

'Population, when unchecked, increases in geometrical ratio. Subsistence increases only in an arithmetical ratio. A slight acquaintance with numbers will shew the immensity of the first power in comparison of the second.

By that law of our nature which makes food necessary to the life of man, the effects of these two unequal powers must be kept equal.

This implies a strong and constantly operating check on population from the difficulty of subsistence. This difficulty must fall some where and must necessarily be severely felt by a large portion of mankind.'

An Essay on the Principle of Population, 1798
T. R. Malthus. 1766–1834.

THE EVOLUTION OF THE ECOSYSTEM

In the preceding chapters we have seen how the hominids evolved from a tropical, arboreal species into a terrestrial, bipedal form that colonized every major biome of the world. In the course of radiating from the tropics, humankind has exerted an increasingly profound influence on the environment inhabited, especially in the last few thousand years when technological proficiency has so rapidly increased. When human numbers were small and population densities were low, humankind adapted both biologically and behaviourally to existing ecosystems without extensive modification of the system's structure. Thus, for the major portion of human prehistory, it has been possible to delineate distinct kinds of adaptations and different ecosystems in which the human species participated; tropical, temperate woodland, grassland, boreal forest, and so on. Today, however, because of the present level of technological development and its associated enormous increase in human population, the boundaries between the various systems have become less significant, and today humankind can be seen to participate in what is essentially a single worldwide ecosystem: the biosphere. Our survival as a species is dependent upon our recognition of this fact.

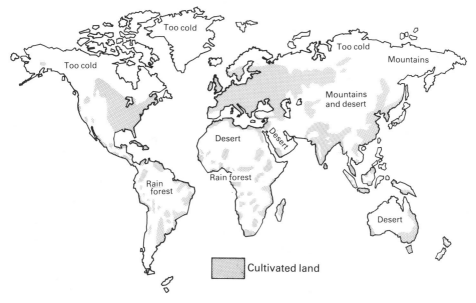

Fig 11.1   Today, all the land that can be cultivated with limited capital investment
is under the plough. To extend cultivation would be extremely expensive, and in
many cases ecologically disastrous. The only remaining resource which might still
yield further food for humans is the ocean. The investment required, however,
may be uneconomic in terms of the possible return. There has never been and
there is not at present a shortage of expensive food: the critical factor is the high
cost of production and transportation which limits the distribution of many foods
to the poorer nations.

In the following pages we shall attempt to pinpoint the critical
systemic variables and the changes in their inter-relationship that have
transformed the human species from a series of relatively isolated small
groups, living in fairly stable relationship with their environment, to a
single, complexly inter-related population whose very means of growth
threatens the viability of the species. We shall attempt to survey our
present problems and our possible future as a species.

Bipedalism and tool-making, together with hunting and food gathering and the demands they made on the early hominids, appear to have
played a critical role in producing those kinds of behaviour we think of
as peculiar to humankind: language, a vastly increased potential for
post-natal learning, cooperation among males, close and prolonged
mother–infant ties that often continue into adult life, and the nuclear
family as the basic economic unit.

The behavioural developments and technological abilities of humankind developed throughout the Early and Middle Pleistocene at a fairly

slow rate. With the controlled use of fire, together with the development of simple facilities such as hide clothing and structures for shelter, humans moved out of the tropics and semi-tropics into temperate and even boreal environments. They still depended almost exclusively on terrestrial resources—plant foods and herbivores—that shared the habitat with them; that is, 'earned' resources. The entire population of humans during the mid-Pleistocene has been estimated to total one million, occurring at a mean density of only ·03 individuals per square mile. Humans lived in groups that probably averaged 25-50 individuals, and their numbers and technological skills were such that their impact on other life forms around them was not highly destructive: they participated in essentially stable ecosystems. The basis for a radical departure from this pattern occurred many millennia later and involved a shift in hunting patterns that profoundly influenced population growth.

By 25 000 BC fully modern humans inhabited most of the Old World and had begun to populate the New World. Everywhere that sites from this period are known, they reveal a common adaptation: the systematic exploitation of migratory herd mammals, with heavy reliance on one or two species[1]. Many sites cluster along valleys that served as migration routes for the herds moving between summer and winter feeding grounds. Rather than following herds about in their annual movements, people settled along these valleys and harvested the animals as they moved through. Although we do not know at what period of prehistory this pattern of hunting first appeared, there is evidence to suggest that it was well established in some areas as early as about 45 000 BP. Certainly by the time we find Upper Palaeolithic sites to have been widely distributed, this was the dominant mode of hunting.

This kind of hunting involved the exploitation of 'unearned' resources and allowed human populations to extract energy from areas far beyond those they occupied. It allowed humans a much higher degree of sedentism, thereby removing much of the necessity for strict limitation of family size. By the late Magdalenian, another kind of 'unearned' resource can be seen to have played a critical role in allowing an even higher degree of sedentism and even further population expansion: riverine resources. The addition of migratory fish, fowl, and aquatic mammals to the diet of late Palaeolithic hunters helped to obviate the lean times of early spring, and after these elements regularly appear in the middens of late Magdalenian sites, there is evidence of rapid population growth. Further, this particular complex of riverine resources appears in archaeological sites in wide areas of the Old and New Worlds at the close of the Pleistocene.

As populations grew rapidly, those locations that were favourable for the exploitation of both terrestrial and aquatic resources became inhabited, and many groups that budded off from parent populations were then forced into less favoured, semi-arid areas. It was in this context that the harvesting of storable cereal seeds became adaptive. The earliest documented attempts at agriculture appear in zones adjacent to established riverside settlements. With the growth of agriculture and its associated stored surplus, there was an increased human dependence upon other kinds of resources: water for irrigation, fuel, and other raw materials, which were transported over wide distances into growing villages and towns.

TRENDS IN HUMAN ECOLOGY

With the development of urbanization which swiftly followed agriculture and its later consequence, industrialization, present-day problems of over-exploitation of natural resources, over-population, and pollution of our environment were already upon us. From our long view of the history of the human species on this planet, there are several trends that can be discerned in tracing our path from that of simple hunter-gatherers to the complicated and troubled human societies that exist today.

The most basic trend, from the earliest Pleistocene camps at the base of Olduvai Gorge to modern-day London or New York, is the increase in the harnessing of free energy from the environment. As a Chinese sage (Tang Zhong) wrote: 'Nature provides the horsepower: mankind provides the harness'. Such energy capture has to do not only with food and its consumption but also with the use of fuels and with harnessing energy sources such as rivers, wind, and so forth. The important distinction between implements and facilities is critical. For the vast range of earlier Pleistocene time, humankind depended upon an essentially implement-based technology. A very marked increase in dependence upon facilities (dams, weirs, houses, irrigation canals, containers, etc.) characterized both terminal Pleistocene and post-Pleistocene adaptations and involved energy control. Machines, which are complex arrangements of implements and facilities, appear relatively late in prehistory; the atl-atl and bow and arrow are examples of early machines. However, from the first urban settlements onwards, machines play an increasingly significant role in helping people to capture and harness and transform energy from their environment.

A second trend that we can see in human adaptations is the ability to straddle an increasing number of trophic levels. From the primarily vegetarian diet of the earliest hominids to the complex cuisine enjoyed in high civilizations, this trend is striking. As humans extended their range from the tropics and sub-tropics into temperate and arctic environments, not only did the food chain become longer, but people depended upon plants and consumers of several levels. The advantage of being adapted to diverse kinds of food is that shortages in supply of one kind of food can often be compensated for by plenty in another. The greater the number of alternatives, the greater the flexibility and reliability of the resource base, but the greater the potential for environmental destruction.

A third trend that appears as a corollary of the systematic use of unearned resources is increased dependence on storage and transport facilities. With the development of agriculture, this dependence became highly important. The existence of any large modern city would not be possible without a complex of storage and transport facilities for imports such as food, water, fuel, and other raw materials.

We see the beginning of these three trends in the early stages of human evolution. The controlled use of fire made possible the exploitation of a broader range of foods and fuels, the trophic span increased, and the storage of food was an important factor in temperate zones. But with the harvesting of unearned resources toward the end of the Pleistocene and the resulting sedentism and population expansion, all three trends systematically accelerated, together with cultural evolution. It is the transporting and storing of food and fossil fuels that has made the recent dramatic development of technology possible. The survival of civilization clearly depends on the maintenance of permanent energy sources (such as atomic, solar, or tidal energy) when world oil supplies run low, as they inevitably will in the coming century. (Table 11.1)

The survival of the human species, however, does not depend on the *further* development of these trends. Because of the quantities of food and fuel that human populations require, further development of them could prove catastrophic.

Table 11.1 Availability of resources

Inexhaustible resources	Exhaustible but renewable resources	Exhaustible and irreplaceable resources

Total amounts of:
  Atmosphere
  Water
  Rock
  Solar energy
	Water in usable condition
	Vegetation
	Animal life
	Human populations
	Certain soil minerals
	Uncontaminated CO_2 and O_2
	Certain ecosystems
		Soil
		Certain minerals
		Rare species
		Certain ecosystems
		Landscape in natural condition
		Much of the ground water supply

RESOURCES AND POPULATION

The rate at which humankind is altering the environment is now reaching the point at which we may seriously and irreversibly damage our prospect for survival[2-3]. Whatever area of human resource extraction activities we investigate, the story is the same, in agriculture, fisheries, forestry, or fuels: productivity is strictly limited or actually

Fig 11.2 This map gives an indication of world population densities. The greatest human settlements lie in north temperate and sub-tropical zones and occur in areas of high agricultural productivity. High population density is characteristic of civilization, but brings with it many ills, including disease and high transportation costs. The important requirement is that the density be related to the resource base. The wealth of a society is best considered in terms of the wealth per head of the population, not the total wealth which may be spread intolerably thinly among starving masses.

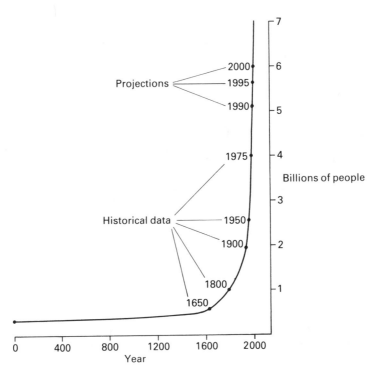

Fig 11.3   The world population curve for humankind is a striking indication of the expansion of human resource extraction and the resulting increase in human biomass. Although projections of 6 billion people by the end of the century are common, it is hard to see how the resources can be generated to support such a number. The quality of life can only fall drastically for all but the richest nations with stable populations.

falling. Even fresh water presents a growing problem: 148 of the largest water basins in the world are controlled by only two countries and 52 more by ten other nations. Population projections imply that water requirements in the year 2000 could be double what they were in 1971 and more will be required if living standards are improved. Competition for control of limited water resources could easily exacerbate international tension.

Improved living standards and increasing population are the twin threats to humankind's survival. What is the population picture today, and what are the prospects?

The present trend towards a later age of marriage and the tendency to reduce family size has meant that the birthrate has been falling in many (but not all) countries during the last decade—both developed and under-developed. Although this is an important and most desirable

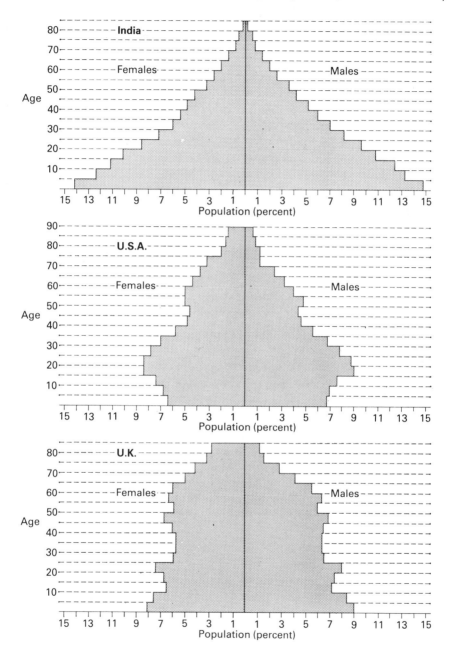

Fig 11.4(a)   Population pyramids indicate short-term population trends very
clearly. In the UK, the population has oscillated a little but is more or less stable,
and in the USA it is moving towards stability. In India a steady expansion is still
underway which can only have disastrous consequences.

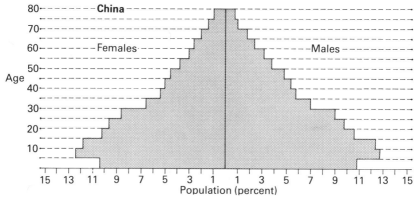

Fig 11.4(b) In China, a reduction in the number of babies born during the last few years is a welcome indication of population control. Every year that such control is delayed in other countries will result in millions of child deaths.

trend, the population of the world is still increasing very rapidly. Recent United Nations projections show that the world's population in the year AD 2000 may be 6000 million $\pm 6\%$. This is 50% higher than the 4000 million people which were living in 1978. Present trends suggest that growth would not stop until AD 2070, with a world population of 11 000 million (Fig 11.3)[4].

These projections are made on the assumption that nutrition and medical aid will continue at present levels per capita. It is clear, however, that the limits to the productivity of the Earth and the social and biological stresses generated by these vast populations will not permit the increase to develop as far as it is projected. The increase will be limited either by planned control of birth, or, more probably, by famine and war. Birth control is spreading slowly in most developed and some underdeveloped countries, but there is still much to be done. China is a remarkable example of a country which has undergone a very rapid population increase but which is at last showing signs of containment. When we examine the population 'pyramids' of (for example) UK, USA, India, and China, we can see some highly significant differences (Fig 11.4). The UK pyramid is that of a relatively stable population where low birth rate and low mortality give the pyramid relatively parallel sides. Compare India where the younger the age group (or cohort), the larger it is. In comparison, China shows a healthy trend: the 0–5 year cohort is smaller than the two which preceded it. Some control of the population, which now numbers 980 million, can therefore be detected—but zero population growth is a long way off. If it could be reached in twenty years, the total population of China would then be 1200 million people.

Both China's and India's population has doubled within the last forty years. In India, the birth rate is still so high that every month 100 000 children die of malnutrition[5]. The drive to develop voluntary (and even compulsory) birth control has been ineffective. The population pyramid shows a very broad base and a very narrow top—the typical pattern of an undeveloped country.

During the last decade, the world's poorest countries have become poorer. Although their gross domestic productivity did increase (by a mere 0·7%), their population increase vastly outweighed and negated this advance. Today there are over 500 million people suffering from gross malnutrition. Poverty, a high birth rate, a high death rate with low life expectancy, go hand in hand with a very bleak future. The human condition in much of the world is deteriorating. There is no doubt that the world ecosystem is becoming destabilized by the very rapid growth of human populations. What were the stabilizing factors in the past?

It is now clear that animal populations are in many known instances adapted to the level of their food resources (or other limiting factors) by a variety of built-in biological mechanisms rather than merely by starvation or thirst, and these come into play in response to signals of incipient overcrowding in advance of serious resource shortages. The signals take the form of the animal's perception of some factor in its environment and they trigger off physiological mechanisms that reduce either fertility, or the survival rate of the foetuses or neonates.

It might be expected that some of these mechanisms might survive during human evolution, and there is indeed no *a priori* reason why they should be expected to be reduced in effectiveness. In fact, there is evidence that such mechanisms probably do exist in human populations though the linkage between cause and effect is often difficult to demonstrate[6]. Miscarriage and infant mortality have been shown in a number of instances to be associated with stress upon the mother during or perhaps prior to pregnancy. The stress will be the product of either food shortages, overcrowding, or social disharmony which may in turn be related to problems associated with resource availability. The most unfavourable conditions induce sterility or stillbirth; somewhat less harsh conditions result in different kinds of impairment of the young which reduce their chance of survival and in an earlier age would probably have resulted in their death. More moderate hardships such as malnutrition, crowding, or the stress of war seem to affect the intelligence, vigour, and motivation of the young. One of the most striking products of population stress is increased irritability and intolerance, so that babies (as well as wives) may be battered, sometimes to death. Some of the problems we find among teenagers are also possibly a stress response which we seem powerless to cure.

Modern humankind has, of course, reduced the effect of these natural checks on populations by building sound-proof apartments and all sorts of architectural and structural devices to reduce the stress effects of over-dense populations. At the other end of the causal chain, the medical profession attempts to alleviate the results of stress by the use of drugs and psychotherapy, and preserves the lives of infants born catastrophically inadequate to cope with normal life. We have said that the causal chain between stress and trauma is not by any means always clearly demonstrated, but that it broadly exists and finds its basis in a natural physiological response which is biologically adaptive is not in question.

Natural selection does not generate happiness or contentment unless it has survival value. If our survival depends on miscarriages, high infant mortality, mental deficiency, criminality, sexual perversion, and so on, they will remain with us inspite of the tremendous efforts of the medical professions. But there is an alternative to hand: cultural methods of population control have been known for thousands of years, and though infanticide and even abortion may seem cruel, they are surely better than the biological responses to overpopulation.

As sedentary populations increased in size and density towards the end of the Pleistocene period, the effects of disease (which perhaps were not so important in the earlier stages of human evolution) became a significant factor in population growth. When, after the development of agriculture, population levels came close to stability again, far more people would have been dying slowly and painfully of disease or malnutrition than previously unless some rational means of population control had been developed.

It is now becoming clear that infanticide has been a very widespread, probably worldwide means of controlling population growth among hunter-gatherers, agriculturalists, and even city dwellers. It was common in Europe in mediaeval times and survived in most parts of the world until Western colonists and missionaries succeeded in stopping it. The distinction of infanticide from murder is still maintained in British law*: if carried out by the mother it is considered comparable to nothing more serious than manslaughter.

* For the purposes of the law an infant is a child under one year old.

In terms of human suffering, it is possibly the best solution to an extravagant birth rate, where modern methods of contraception are not known. Some peoples in the past have maintained a very enlightened view of this problem. In the Ellice Islands of the Pacific, for example, infanticide was ordered by law; only two children were allowed to a family, as the islanders were afraid of the scarcity of food[7].

The well-meaning interference in stable human societies perpetrated by missionaries, both medical and religious, has had a disastrous result on the lives of those peoples whose populations have increased faster than their ability to develop their resource base. The result has been the appearance of poverty-stricken 'underdeveloped' countries. It is more accurate to state that there are no poor countries—only over-populated ones. Because the resource base of each country varies, the carrying capacity of each country varies. In an ideal world, the population of a nation should accurately reflect its resource base. *This is the only way that the standard of living of people in the 'poorer' countries can ultimately be raised in a finite world ecosystem.* Without the control and even (in some instances) the reduction of population size, per capita productivity can no longer be significantly increased.

The situation in the United Kingdom is instructive. With a present population of about 56 million, we produce only 54% of our food requirements. The rest of our food must be paid for by exports of goods and services. So long as we can generate these exports and sell them, we can support this relatively vast population, but we are frighteningly dependent upon world markets—upon the world's need for what we produce—and there is no guarantee of that.

Interdependence through trade is humankind's equivalent to the energy flow in a natural ecosystem and as such is productive and stabilizing. But interdependence requires stability in trade-patterns for the survival of all components of the system. If we are to survive, we require stability in all parameters: population size, resource base, and productivity—at levels which the Earth can maintain indefinitely[8].

It was at the end of the nineteenth century when new discoveries in medicine (such as Pasteur's demonstration of micro-organisms as the cause of disease) brought about a new acceleration in population size, not by increasing the number of babies allowed to survive, but by reducing the death rate. Life expectancy at birth is today over seventy years in western countries, but in many parts of Africa and Asia, it may be as low as under forty-five years. Disease is therefore still an important factor in population control, especially in underdeveloped countries where there is either no western medicine or doctors are unable to obtain the necessary drugs to treat their patients. In many areas, however, we are in a position to preserve lives only to suffer malnutrition and starvation. In all, about 1500 million people are undernourished or malnourished and, as we have seen, 500 million are either chronically hungry or starving.

The production of more children than the number for whom the parents can guarantee a food supply does not today threaten most parents' survival in any immediate sense in most countries (as it does and did among nomadic hunter-gatherers)[9]. Family limitation thus becomes no longer an act vital to the parents' survival, so that family size in many countries is still determined by means of disease and starvation. It is fortunate that methods of birth control are available today which allow us to limit family size without resorting even to abortion or infanticide. Primitive techniques of birth control were known to the ancient Greeks and Egyptians, but only in the last twenty years has the medical profession developed means of control which are simple to use and effective. This is one of the few instances (another is perhaps the discovery of anaesthetics) where scientific and technological progress can be, and occasionally is, of unquestionable benefit to humankind. The need to limit the birth rate is paramount. The cost to the environment of any further increase (and indeed of maintaining even our present numbers) will be incaculable.

DEVELOPMENT AND EXPLOITATION

In the absence of effective worldwide birth control, the only alternative humanitarian course of action is to attempt to increase food supplies and other essential resources. The history of the last 10 000 years is the story of agricultural development and it is still possible today to increase productivity by the use of modern farming methods, modern technology, newly bred varieties of plants and animals, and ample organic or inorganic fertilizer. We have seen enormous progress in agricultural techniques and the population which the world can support today is probably at least 100 times as great as it was at the time of the birth of agriculture. This continuous and enriching development of agriculture and animal husbandry led Voltaire to voice what, in the eighteenth century, seemed so obvious: 'Nature is inexhaustible, and untiring labour is a god that rejuvenates her.'

We now know, alas, that nature is finite in her beneficence, and that

Fig 11.5   The Aswan dam is one of the most important and expensive irrigation developments ever undertaken. It brought into cultivation thousands of acres in Upper Egypt, but its effect on the whole Nile ecosystem is giving cause for serious concern.

we have come close to her limits. The underdeveloped countries cry out to the West to help them to achieve further 'development'—a word which apparently means not only agricultural development but industrial development and the construction of roads, railways, and so on—known as infra-structure. The expressed aim of such development is to better the lives of the people.

Today 'development' is in most instances a euphemism for environmental exploitation. Most parts of the world capable of sustaining a high level of agricultural productivity are already developed—those areas not yet developed (such as the Amazon basin) are probably not suitable for it. So many apparently valuable development projects, such as the drilling of bore-holes in Masailand and the vast irrigation schemes in the Punjab and in California, have had appalling and unexpected results. Many more examples could be quoted where deserts and disease have spread, and where industry has polluted vital water sources. The building of the great Aswan dam across the Nile in Upper Egypt is probably the most recent and striking example of a supposedly wonderful development project which is proving to have devastating side effects (only some of which were foreseen by ecologists) (Fig 11.5). Most of these effects result from the fact that the Nile's huge annual load of fertile silt which was previously carried down to enrich the farmlands of the valley and delta, as well as the Mediterranean, is now settling out above the dam in Lake Nasser. As a result, not only will the lake eventually silt up, but large quantities of expensive artificial fertilizers have to be spread on the once naturally fertilized farmland. A sardine fishery off the mouth of the Nile has collapsed, and fish catches are dropping throughout the Eastern Mediterranean.

Furthermore, the rate of evaporation on Lake Nasser is now such that the total amount of water in the Nile has been reduced, and Egypt is short of water. The recently dug irrigation schemes carry waterborne schistosomiasis (blood fluke disease) to millions of people; and as the water table rises, salinity has developed in farmland. And this is only part of the story—but it is enough to demonstrate that technology and good intentions, properly funded, are not enough.

The tangle of problems in energy, population, and food that confront us with such unexpected complexity have given rise to the expression 'world problematique'[10]. We face an increasing complexity of problems with a level of understanding and capability which lags far behind. We have to stop changing and start to study the world, and then try to educate humanity before it is too late, before irreversible degradation and pollution become any more widespread. We are close to the limits which our environment necessarily imposes. It is certainly impossible to increase food production further in many areas, and elsewhere the cost is increasing, both socially and environmentally. In many countries the high costs of development already outweigh the returns which we may expect to gain[11].

If we were to stabilize our population at a level the biosphere could support, then we could turn our efforts, not to maximize our place in nature, but to optimize it. We no longer need to think in terms of quantity, but of quality, for it is to improve the quality of the human environment and of human lives that is the proper aim of humankind.

THE MANIFOLD AND THE ONE

The study of ecology teaches us of the interdependence of all parts of the planet Earth in systemic relationship: the geophysical substrate, the atmosphere and climate, the plants and animals. It is also obvious that the Earth depends on the Sun for its energy source and the Moon for its tides: the system is an open system and part of the cosmos. Because of this total interdependence of all the myriad components of the whole, it is by no means far-fetched to compare our entire world system with an individual organism. We accept the systemic nature of an individual for we know that there is an obvious interdependence of the different organs. If we see the whole planet in this way, we will hesitate to make major and fundamental changes to particular components rapidly and thoughtlessly.

The analytic reductionist thinker, who examines one part of the system only to see how it affects another part, looks for straight lines of cause and effect: A→B→C. Such a thinker believes that if you do something that is good and productive, then more of the same will necessarily be better. This simple logic is dangerous, because it is based on the entirely false assumption that causality in nature is linear rather than systemic. The thinker who recognizes the complex web of systemic interaction (e.g. Fig 1.8), will, in contrast, expect that if you make major changes—if you do more of the same good, the local situation may improve for a while, but then will probably get worse. The system will adjust. Such a view recognizes the profound truth that there is an *optimal* value for everything (which is not maximal), whether it is the size of a farm, a company, or a city, and that major alterations in the values of the components of a system will reverberate throughout the system with extensive and probably unpredictable and destructive results. This is indeed our experience. The systemic model of causality is not as neat and tidy as the linear one, but as we have seen, it comes very much closer to describing the nature of our living planet.

A people's religion, their belief system, both reflects and to a very great extent determines their attitude to the natural world. We continually act on the basis of our beliefs: the Judaeo-Christian concept of the conquest of nature has had a devastating effect on our planet. Clearly it is absolutely imperative that we replace it by a belief much more subtle, which reflects the truth of our present predicament, and which will ultimately be much more rewarding for humanity as a whole. We must come to believe ourselves to be what indeed we are— part of the intricate and balanced structure of the natural world; not a conqueror, who bends nature to his will and exploits her wealth.

Therefore it is no longer a mysterious paradox to see nature as both the manifold and the one. The components of the natural world are myriad, but they constitute a single living system. There is no escape from our interdependence with nature; we are woven into the closest relationship with the Earth, the sea, the air, the seasons, the animals and all the fruits of the Earth. What affects one affects all—we are part of a greater whole—the body of the planet. We must respect, preserve, and love its manifold expression if we hope to survive.

References

1   Binford, L. R. 1968 Post-Pleistocene Adaptations, in *New Perspectives in Archaeology*, S. R. Binford and L. R. Binford, eds. (Chicago: Aldine Publishing Co.).

2   Ehrlich, P. R. and A. H. Ehrlich 1977 *Ecoscience: Population, Resources, Environment* (San Francisco: Freeman).

3   Hutchinson, G. E. et al. 1970 eleven articles in *Scientific American* 223 (3): 44–208.

4   Dadzie, K. K. S. et al. 1980 ten articles in *Scientific American* 243 (3): 54–181

5   Kendall, M. 1980 *Population Reports, World Fertility Survey*. Johns Hopkins University.

6   Stott, D. H. 1969 Cultural and Natural Checks on Population Growth, in *Environment and Cultural Behaviour*, A. P. Vayda, ed. (Natural History Press: New York).

7   Wilkinson, R. G. 1973 *Poverty and Progress* (Praeger: New York).

8   Odum, E. P. 1960 The Strategy of Ecosystem Development. *Science*, N.Y. 164: 262–70.

9   Hardin, G. 1969 *Population, Evolution and Birth Control* (San Francisco: Freeman).

10   Anon. 1980a *Approaching the Twentyfirst Century*. US Government Publs. Washington.

11   Anon. 1980b *World Conservation Strategy*. World Wildlife Fund/International Union for the Conservation of Nature.

Suggestions for further reading

Bender, B. 1975 *Farming in Prehistory: from Hunter-gatherer to Food Producer* (London: John Baker).

Brown, L. R. and Finsterbusch, G. W. 1972 *Man and his Environment: Food* (London: Harper & Row).

Campbell, B. G. 1982 *Humankind Emerging* (Boston: Little Brown & Co.).

Hoagland, H. 1964 Mechanisms of Population Control. *Daedalus* 93: 812–829.

Odum, E. P. 1971 *Fundamentals of Ecology* (Philadelphia & London: W. B. Saunders).

Pfeiffer, J. 1969 *The Emergence of Man* (New York: Harper & Row).

Service, E. R. 1962 Primitive Social Organization: an evolutionary perspective (New York: Random House). 1966 *The Hunters* (Englewood Cliffs: Prentice Hall Inc.).

Weiner, J. S. 1971 *The Natural History of Man* (London: Weidenfeld).

Index